Lecture Notes in Control and Information Sciences

Edited by M. Thoma

For information about Vols. 1– 42 please contact your bookseller or Springer-Verlag.

Lecture Notes in Control and Information Sciences

Edited by M. Thoma and A. Wyner

115

H. J. Zwart

Geometric Theory for Infinit Dimensional Systems

Springer-Verlag
Berlin Heidelberg GmbH

Author
Hans J. Zwart
Faculty of Applied Mathematics
University of Twente
P. O. Box 217
7500 AE Enschede
The Netherlands

ISBN 978-3-540-50512-9   ISBN 978-3-540-46026-8 (eBook)
DOI 10.1007/978-3-540-46026-8

2161/3020-543210

Geometry may sometimes appear to take the lead over analysis but in fact precedes it only as a servant goes before the master to clear the path and light him on his way.

James Joseph Sylvester

PREFACE AND ACKNOWLEDGEMENT

In the spring of 1984 I started with my research on geometric theory for infinite dimensional systems. The research topic was suggested to me by Ruth Curtain, who had done some preliminary investigations on this topic. Many questions were at that time still open and a more fundamental theory was still missing. We knew that the key–concept in geometric theory for finite dimensional systems, that is (A,B)–invariance, has lost its strength for infinite dimensional systems. So I began to look for different concepts which would be more appropriate for infinite dimensional systems. It turned out that these were the concepts of open–loop invariance and frequency invariance. Although the concept of frequency invariance had already been introduced for finite dimensional systems by Hautus, he did not give it any special name. I have chosen this name, since this expresses in a concise way that this is an invariance concept in the frequency domain. Once the equivalence between open–loop, frequency, and closed loop invariance was established, the solvability of various disturbance decoupling problems came relatively easy. In this monograph three disturbance decoupling problems are studied: the Disturbance Decoupling Problem (DDP), the Disturbance Decoupling Problem with Measurement Feedback (DDPM) and the Disturbance Decoupling Problem with Measurement Feedback and Stability (DDPMS). The theory can easily be extended to other disturbance decoupling problems, with the notable exception of the 'almost' version, which are studied in the finite dimensional case by Willems and Trentelman, see e.g. [39]. The theory for the almost disturbance decoupling problems is one of the main still open problems in geometric theory for infinite dimensional systems.

The monograph is addressed to researchers in the field of geometric theory of infinite dimensional systems. In this book I shall use basic concepts of the infinite dimensional system theory as C_0–semigroup, approximate controllability, initial observability, which are covered in the second and third chapter of Curtain and Pritchard [9]. This book is self–contained with respect to the notions of the geometric theory, although sometimes we shall refer to the references for the finite dimensional case.

Although it may seem that writing a monograph and doing research is a solo occupation, in reality it is a team occupation and I owe my team members of the Groningen System Theory Group a great debt of gratitude.

First of all I want to thank Ruth Curtain who found always the time and the patience to listen to my ideas. Her guidance made sure that my research would not wander off in queer directions.

During the past four years it has been a great pleasure to share the office with Jan Bontsema. As a room-mate he always had a lending ear to listen to my (sometimes obscure) problems and his relativizing way of looking at these problems really meant a lot to me. Furthermore I would like to thank the other members of the System Theory Group in Groningen; Harry Croon, Christiaan Hey, Hans Nieuwenhuis, Paula Rocha, Siep Weiland and Jan Willems, for the privilege of working with them. They all contributed in their own way to this research and made our lunch breaks a very cosy hour. I also express my gratitude to Hans Schumacher whose insight into the problem plus his remarks and ideas helped me to get my research started.

Special thanks go to Erik Thomas, Malo Hautus and Luciano Pandolfi for the careful way they read this monograph. Their discussions and interest from different mathematical backgrounds all contributed to this research.

I also want to thank the office of the mathematics department of the university of Groningen for their help during the last years. Special thanks go to Janieta Schlukebir for typing part of this monograph.

This research was sponsored by the Netherlands Organization for Scientific Research (N.W.O.), under grant no. 10-64-06 for which I am grateful.

Groningen,   Hans Zwart
July, 1988

CONTENTS

INTRODUCTION

The aim of this monograph is to present a geometric approach to disturbance decoupling problems for infinite dimensional systems. Before we go into details we shall give an outline of the disturbance decoupling problems.

By a disturbance decoupling problem we mean a problem of the following type. Let Σ denote a system for which we can distinguish two classes of inputs, $u(.)$ and $q(.)$, and two classes of outputs, $y(.)$ and $z(.)$, as is schematized in figure 1 by a signal flow graph.

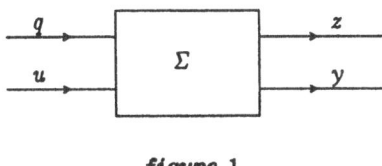

figure 1

In this system we regard the input $q(.)$ as an undesired influence (disturbance) on the system. In general this input will influence both outputs. Now a disturbance decoupling problem amounts to constructing a second system Σ_f, which takes as input $y(.)$ and gives as output a control input $u(.)$, such that $z(.)$ has become independent of the disturbance input $q(.)$. Pictorially we have

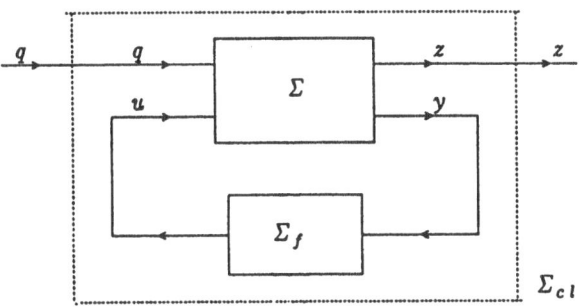

figure 2

So if we regard the interconnection of the systems, Σ and Σ_f, as one system, Σ_{cl}, as schematized in figure 2, then we see the input $q(.)$ as a disturbance signal which should not influence the output $z(.)$, and we use the measurement $y(.)$ in order to design a control input $u(.)$ which cancels the effect of the disturbance $q(.)$ on $z(.)$, i.e. which decouples $z(.)$ from $q(.)$. Therefore we shall call $u(.)$ the control input, $q(.)$ the disturbance input, $y(.)$ the measured output or measurement and $z(.)$ the to be decoupled output. The next example shows that this problem is solvable for some systems Σ.

Example 1.

In this example we consider the binary distilation column as studied by Takamatsu, Hashimoto and Nakai [38]. The system is assumed to be in an equilibrium and we want to make the composition of the distilate and the composition of the reboiler independent of changes in the composition of the feed stream. As a model for this distilation column we take

(1) Σ $\quad \begin{cases} \dot{x}(t) = Ax(t) + Bu(t) + Eq(t) \\ y(t) = x(t) \\ z(t) = Cx(t) \end{cases}$

where

$x(.) = \left(x_1(.), x_2(.), .., x_{11}(.)\right)^t$, $\quad x_i(.)$ denotes the difference between the liquid composition on the i–th tray and its equilibrium value,

$u(.) = \left(u_1(.), u_2(.)\right)^t$; $u_1(.)$ is the difference of the flow rate of the liquid stream and its equilibrium value, $u_2(.)$ is the same, but now for the vapor stream,

$z(.) = \left(z_1(.), z_2(.)\right)^t$; $z_1(.) = x_1(.)$, $z_2(.) = x_{11}(.)$.

Furthermore

$A = (a_{i,j})$ is a tri–diagonal matrix, of which the upper diagonal, the diagonal and the lower diagonal are given by respectively,

$\{a_{j,j+1}\} = (0.105, 0.469, 0.529, 0.596, 0.569, 0.718, 0.799, 0.901, 1.021, 1.142)$,

$\{a_{j,j}\} = -(0.174, 0.943, 0.991, 1.051, 1.118, 1.584, 1.64, 1.721, 1.823, 1.943, 0.171)$,

$\{a_{j+1,j}\} = (0.522, 0.522, 0.522, 0.522, 0.522, 0.922, 0.922, 0.922, 0.922, 0.115)$,

B is given by

$$B = 10^{-5} * \begin{bmatrix} 0 & 328 & 384 & 400 & 376 & 308 & 136 & 288 & 308 & 300 & 32 \\ 0 & -244 & -288 & -304 & -280 & -232 & -312 & -382 & -412 & -396 & -42 \end{bmatrix}^t$$

$$C = \begin{bmatrix} 1 & 0 & 0 & 0 & 0 & 0 & 0 & 0 & 0 & 0 & 0 \\ 0 & 0 & 0 & 0 & 0 & 0 & 0 & 0 & 0 & 0 & 1 \end{bmatrix} \quad \text{and } E = \left(0\ 0\ 0\ 0\ 0\ 0.4\ 0\ 0\ 0\ 0\ 0 \right)^t.$$

It is shown in [35] that there exists a simple system Σ_f which makes the output $z(.)$ independent of the disturbances. This system is given by

(2) Σ_f: $u(t) = Fx(t)$,

where F is given by, (correct to five digits)

$$F = \begin{bmatrix} 0 & 0 & -330.06 & 0 & 0 & 0 & 0 & 0 & 0 & 470.17 & 0 \\ 0 & 0 & -251.47 & 0 & 0 & 0 & 0 & 0 & 0 & 632.04 & 0 \end{bmatrix}.$$

So $u_1(.) = -330.06*x_3(.) + 470.17*x_{10}(.)$ and $u_2(.) = -251.47*x_3(.) + 632.04*x_{10}(.)$.

In figure 3 the output $z(.)$ is drawn for the system Σ and for the system Σ_{cl}. In both systems the same disturbance signal was used.

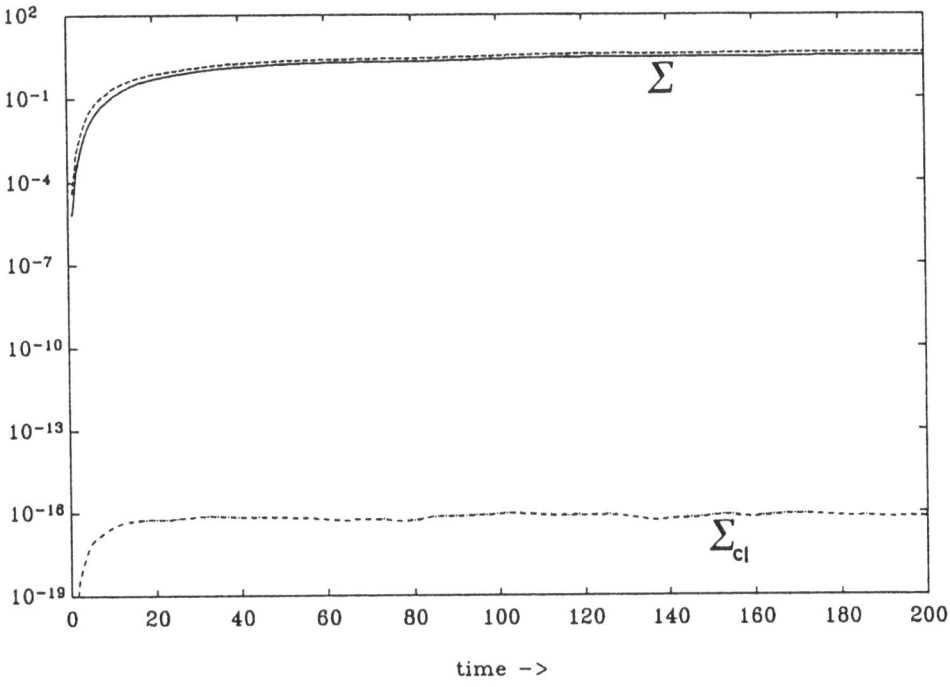

time $->$

figure 3

Note that for the system Σ_{cl} the output signal has become a factor 10^{16} smaller and is of the same order as the machine accuracy. So in this example we can reject the effect of disturbances   □

A natural question now arises: under which conditions is disturbance decoupling possible and how can one construct the feedback system Σ_f? Before we can solve this question we should specify which systems we want to consider. A simple, but not unimportant, class of systems is the class of linear time–invariant, finite dimensional systems. For this class we shall give the solution of the disturbance decoupling problem as presented by Basile & Marro [1] and Wonham [42].

Disturbance Decoupling Problem for Finite Dimensional Systems

The class of systems that we shall consider in this section is the class of systems that have the following representation

(3)
$$\dot{x}(t) = Ax(t) + Bu(t) + Eq(t);$$
$$y(t) = Cx(t);$$
$$z(t) = Dx(t); \quad t \geq 0; \quad x(0) = x_0$$

where $x(t)$, $u(t)$, $q(t)$, $y(t)$ and $z(t)$ are time trajectories in respectively R^n, R^m, R^q, R^p and R^z, and A, B, C, D and E are matrices of appropriate size. Notice that the system of example 1 is of this class.

In order to obtain some insight into the disturbance decoupling problem we assume that we measure the full state of the system, i.e. we assume that $y(t) = x(t)$, or equivalently $C = I_n$. Furthermore we want the system Σ_f to be as simple as possible, and so we assume it to be time invariant and memoryless. So we assume that

(4) Σ_f: $u(t) = Fx(t)$

with F a matrix. Or in other words, $u(t)$ is a linear combination of the components of the state at time t. The problem at hand is to find a feedback system (4) such that after interconnecting (3) and (4) we have that the disturbance input $q(.)$ has no influence on the output $z(.)$ for all disturbance signals. This problem is commonly known as <u>the</u> disturbance decoupling problem.

Definition: Disturbance Decoupling Problem

The Disturbance Decoupling Problem is to find, if possible, for the system (3) a feedback system of the form (4) such that in the closed loop system the disturbance input $q(.)$ has no influence on $z(.)$ for all disturbance signals.

It is standard to refer to this problem by its initials, DDP, and we shall continue this tradition.

For the class of systems defined in (3) we can give a precise formula for the closed loop behaviour of the system. This solution is given by the well–known variation of constant formula,

$$(5) \qquad z(t) = De^{(A+BF)t}x_0 + \int_0^t De^{(A+BF)(t-s)}Eq(s)ds$$

Since we assumed no prior knowledge about the disturbance input $q(.)$, we must have that the function $De^{(A+BF)t}E$ is identically zero on $[0,\infty)$, in order to make $z(.)$ independent of $q(.)$. So DDP is solvable if and only if we can find a feedback law F such that $De^{(A+BF)t}E \equiv 0; \ t \geq 0$.

This last problem is very hard to solve directly. Since one has to calculate $e^{(A+BF)t}$ for all F. However there is a simple necessary condition which we can deduce from it, namely for $t=0$ we have that $De^{(A+BF)0}E = DE$, and this must be zero. So DDP is solvable, only if $DE = 0$. This condition shows in particular that DDP is not solvable for every system in our class. In example 2 we shall see that the condition $DE = 0$ is not sufficient.

We can interpret the condition $De^{(A+BF)t}E \equiv 0; \ t \geq 0$ in the following system theoretic way, namely all trajectories of the system $\dot{x}(.) = (A+BF)x(.)$ that start in $Im\ E$ will remain in the kernel of D. Let V denote the reachable subspace for the system $(A+BF,E)$, where F is the feedback law that solves DDP. This subspace is defined as

$$(6) \qquad V = span\left\{ e^{(A+BF)t}Eq \right\}; \text{ where the span is taken over } t \geq 0 \text{ and } q \in \mathbb{R}^q$$

By the solvability of DDP we must have

$$(7.a) \qquad Im\ E \subset V \subset Ker\ D$$

where $Im\ E$ and $Ker\ D$ denote the image of E and the kernel of D respectively, and the semigroup property of $e^{(A+BF)t}$ implies that V also satisfies

$$(7.b) \qquad e^{(A+BF)t}V \subset V; \ t \geq 0$$

On the other hand, suppose that there exists a subspace $V \subset \mathbb{R}^n$, which for some F satisfies the conditions $(7.a)$ and $(7.b)$. Then DDP is solvable with this F, since $e^{(A+BF)t}Im\ E \subset e^{(A+BF)t}V \subset V \subset Ker\ D$, and thus $De^{(A+BF)t}E \equiv 0$.

So in conclusion we can say that DDP is solvable if and only if there exists a subspace V and a feedback F which satisfy (7.a) and (7.b). One could argue that this result makes the problem more difficult; not only must one construct a feedback F, but also one must construct a subspace $V \subset R^n$ with the properties (7.a) and (7.b). The next theorem which can be found in Basile & Marro [1] and Wonham [42, p. 88] shows that one only has to construct a subspace $V \subset R^n$ with special properties which are easy to check.

Theorem

Let V be a linear subspace of R^n. Then the following properties are equivalent:

i)   There exists an F such that $e^{(A+BF)t} V \subset V$.

ii)   There exists an F such that $(A+BF)V \subset V$.

iii) $AV \subset V + Im\, B$

Some remarks must be made about this theorem; first the feedbacks in i) and ii) can be chosen to be the same, and second the construction of the feedback F from iii) to ii) or i) is done by solving linear equations, which is a relatively easy problem, see example 2.

Before we continue with the DDP we shall briefly refer to invariance concepts related with the system $\dot{x}(t) = Ax(t) + Bu(t)$. Since the solution of the differential equation $\dot{x}(t) = (A+BF)x(t)$; $x(0) = x_0$ is given by $x(t) = e^{(A+BF)t} x_0$, we can replace assertion i) by

i')   There exists an F such that all solutions of the differential equation $\dot{x}(t) = (A+BF)x(t)$ which start in V will remain in V.

We can now pose the question whether one would gain more if one were to allow for 'arbitrary' inputs instead of inputs generated by feedback. In other words are there subspaces V which do not satisfy i'), but do satisfy:

iv)   For every $x_0 \in V$ there exists a continuous input $u(t)$ such that the solution of $\dot{x}(t) = Ax(t) + Bu(t)$; $x(0) = x_0$ remains in V.

The answer to this question is negative. Since if $x(t)$ is in V for all $t \geq 0$, then $\dot{x}(t)$ is in V for all $t \geq 0$. So for $t = 0$ we have that $Ax_0 = Ax(0) = \dot{x}(0) - Bu(0)$ $\in V + Im\, B$. Thus by the equivalence of i') and iii) we have that there exists an F such that the solutions of $\dot{x}(t) = (A+BF)x(t)$; $x(0) = x_0$ remain in V.

From the theorem and the argument above we see that property iii) is of great importance. This property has been given a special name.

Definition: (A,B)-invariance

    A subspace V is (A,B)-invariant if

(8) $\qquad A V \subset V + Im\, B$

So with respect to the DDP we see that this problem is equivalent to finding an (A,B)-invariant subspace V with property (7.a). Let $\Im(A,B;Ker\, D)$ denote the class of all subspaces that are (A,B)-invariant and contained in the kernel of D. Note that we are looking for an element in $\Im(A,B;Ker\, D)$ which contains $Im\, E$.

Trivially the zero subset is an element of $\Im(A,B;Ker\, D)$. Furthermore it is an easy exercise to prove that the sum of two elements in $\Im(A,B;Ker\, D)$ is again an element of $\Im(A,B;Ker\, D)$. So by the finite dimensionality of $Ker\, D \subset R^n$ there will exists a supremal element in $\Im(A,B;Ker\, D)$, which we shall denote by $V^*(Ker\, D)$. Thus for every element V in $\Im(A,B;Ker\, D)$ we have that $V \subset V^*(Ker\, D)$. From this we have the following theorem as an easy corollary.

Theorem

    The DDP is solvable if and only if $Im\, E \subset V^*(Ker\, D)$.

    Furthermore the feedback that solves the DDP can be any feedback F that satisfies $(A+BF)V^*(Ker\, D) \subset V^*(Ker\, D)$.

We have mentioned that calculating the feedback F is a linear problem. Now we shall see that the calculation of $V^*(Ker\, D)$ is also fairly easy. Define the sequence V^μ according to

(9) $\qquad V^0 = Ker\, D; \quad V^\mu = Ker\, D \,\cap\, A^{-1}\!\left(Im\, B + V^{\mu-1}\right); \quad \mu = 1,2,..$

where $A^{-1}(X)$ is the set consisting of all elements y such that $Ay \in X$.

By induction it is easy to show that $V^\mu \subset V^{\mu-1}$, and for some $k \leq \dim(Ker\, D)$ we have that $V^k = V^{k+1} = V^{k+2} = ...$ Furthermore we have that $V^k = V^*(Ker\, D)$.

A precise proof of this can be found in e.g. Wonham [42, p.91].

So we have obtained an algorithm which calculates in finitely many steps the supremal (A,B)-invariant subspace contained in the kernel of D. Some

authors call this algorithm the Invariant Subspace Algorithm (ISA).
Now we illustrate the previous theory by a simple example.

Example 2

Consider the following system on $\mathbb{R}^3$

$$\dot{x}(t) = \begin{bmatrix} 0 & 0 & -24 \\ 1 & 0 & -26 \\ 0 & 1 & -9 \end{bmatrix} x(t) + \begin{bmatrix} -1 \\ 1 \\ 0 \end{bmatrix} u(t) + \begin{bmatrix} 1 \\ 3 \\ 0 \end{bmatrix} q(t);$$

(10)
$$z(t) = \begin{pmatrix} 0 & 0 & 1 \end{pmatrix} x(t)$$

We want to solve the DDP for this system. Note that the necessary condition
of $Im\, E = span\{ \begin{bmatrix} 1 \\ 3 \\ 0 \end{bmatrix} \} \subset \{x_3 = 0\} = Ker\, D$ is satisfied, so to check whether or not
DDP is solvable we have to calculate $V^*(Ker\, D)$. This can be done by means
of algorithm (9), yielding:

$$V^0 = Ker\, D = \{x_3 = 0\},$$

$$V^1 = Ker\, D \cap A^{-1}\{ Ker\, D + Im\, B \} = Ker\, D \cap A^{-1}\{\{x_3 = 0\} + span\{ \begin{bmatrix} -1 \\ 1 \\ 0 \end{bmatrix} \}\}$$

$$= Ker\, D \cap A^{-1}\{ \{x_3 = 0\} \} = Ker\, D \cap \{ \begin{bmatrix} x_1 \\ x_2 \\ x_3 \end{bmatrix} | x_2 - 9x_3 = 0 \} = span\{ \begin{bmatrix} 1 \\ 0 \\ 0 \end{bmatrix} \}$$

Now $V^1 + Im\, B = \{x_3 = 0\}$, so $V^2 = V^1$ and this implies that $V^*(Ker\, D) = V^1 = span\{ \begin{bmatrix} 1 \\ 0 \\ 0 \end{bmatrix} \}$.
So pictorially we have the following situation.

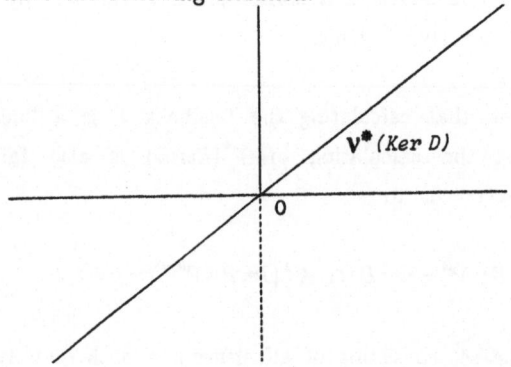

$V^*(Ker\, D)$

0

Since $Im\, E$ is not contained in $V^*(Ker\, D)$ we have that the DDP is not
solvable for this example.
We shall complete this example with the calculation of a F such that
$(A + BF)V^*(Ker\, D) \subset V^*(Ker\, D)$.

Let $v = \begin{pmatrix} x_1 \\ 0 \\ 0 \end{pmatrix}$ be an element of $V^*(Ker\ D)$. Then by the (A,B)–invariance of $V^*(Ker\ D)$ we can decompose Av as an element of $Im\ B$ and $V^*(Ker\ D)$, yielding

(11) $\qquad Av = \begin{pmatrix} 0 & 0 & -24 \\ 1 & 0 & -26 \\ 0 & 1 & -9 \end{pmatrix}\begin{pmatrix} x_1 \\ 0 \\ 0 \end{pmatrix} = \begin{pmatrix} 0 \\ x_1 \\ 0 \end{pmatrix} = \begin{pmatrix} x_1 \\ 0 \\ 0 \end{pmatrix} + x_1\begin{pmatrix} -1 \\ 1 \\ 0 \end{pmatrix}$

If we take for the feedback F any $F = (f_1, f_2, f_3)$ with $f_1 = -1$, then $F\begin{pmatrix} x_1 \\ 0 \\ 0 \end{pmatrix} = -x_1$, and

(12) $\qquad (A+BF)v = Av + BFv = v + x_1 B + BFv = v + x_1\begin{pmatrix} -1 \\ 1 \\ 0 \end{pmatrix} + (-x_1)\begin{pmatrix} -1 \\ 1 \\ 0 \end{pmatrix} = v \in V^*(Ker\ D).$

So any feedback $F = (f_1, f_2, f_3)$ with $f_1 = -1$ satisfies $(A+BF)V^*(Ker\ D) \subset V^*(Ker\ D)$. From equation (12) we also see that 1 is always an eigenvalue of $A+BF$. From this it seems that there exists no feedback which keeps $V^*(Ker\ D)$ invariant and is also stable (all eigenvalues of $A+BF$ in the open left half plane). We shall see in the sequel that this is indeed true.   □

We shall conclude this section with some results of Hautus [19], which give in the frequency domain a ·necessary and sufficient condition for the solvability of the DDP.

The disturbance decoupling problem is to find a feedback F such that $De^{(A+BF)t}E \equiv 0$; $t \geq 0$. By taking the Laplace transform we see that $De^{(A+BF)t}E \equiv 0$; $t \geq 0$ if and only if $D(s - A - BF)^{-1}E \equiv 0$. We can rewrite this equation as

(13) $\qquad 0 \equiv D(s - A - BF)^{-1}E = D(s - A)^{-1}(s - A)(s - A - BF)^{-1}E =$

$\qquad D(s - A)^{-1}(s - A - BF + BF)(s - A - BF)^{-1}E = D(s - A)^{-1}E + D(s - A)^{-1}BF(s - A - BF)^{-1}E.$

Since $F(s - A - BF)^{-1}E$ is a strictly proper rational function we see from (13) that the DDP is solvable only if there exists a strictly proper rational function $X(s)$ which satisfies

(14) $\qquad D(s - A)^{-1}E + D(s - A)^{-1}BX(s) = 0$

It was proved by Hautus [19] that the solvability of (14) by a strictly proper rational function is also sufficient.

Notice that the the functions $D(s-A)^{-1}E$ and $D(s-A)^{-1}B$ are respectively the transfer function of the disturbance to the output and the transfer function of the input to the output. Furthermore we see for one dimensional disturbances, inputs and outputs that a solution of (14) is given by $X(s)=-D(s-A)^{-1}E/D(s-A)^{-1}B$. So if s_0 is a zero of $D(s-A)^{-1}B$ and not a zero of $D(s-A)^{-1}E$, then s_0 is a pole for every solution of (14). Thus if $D(s-A)^{-1}B$ has zeros in the right half plane, then no stable solution of (14) exists, see also example 2.

We close this section with the following remark. In this section we have mainly concentrated on the question whether or not disturbance decoupling was solvable with a system Σ_f of the form (4). It has been shown by Basile & Marro [1] that for the case that one measures the state of the system (3), i.e. $C=I_n$, the feedback system Σ_f can always be chosen of this form. So if disturbance decoupling is solvable for some system Σ_f which takes as input $x(.)$, then there exists a system $\tilde{\Sigma}_f$ of the form (4) which solves the DDP

In the former section we have tried to give an introduction into the ideas behind the geometric approach to disturbance decoupling problems. The disturbance decoupling problem and its relatively easy and nice solution in terms of subspaces which one can calculate marked the starting point of a totally new theory, now known as the geometric theory for finite dimensional systems. In terms of this theory a lot of disturbance decoupling problems were solved, as well as many related problems. Out of this theory arose a deeper insight into the system theoretic structure of the state space. As is to be excepted after almost twenty years there is a huge amount of literature on this subject, see e.g. [1], [2], [19], [33], [34], [35], [36], [39], [40], [42] and [43] and the references there. For an interested reader who is not familiar with the geometric theory the first article of Basile and Marro [1] and the book by Wonham [42] will give a good introduction to this theory.

For the class of finite dimensional systems the solvability of disturbance decoupling problems is very well understood. Inspired by these nice results authors have tried to extend these results to other classes of systems, see

Hautus [20] for systems over rings, Pandolfi [27] for systems described by delay equations and Schmidt & Stern [32] and Curtain [6], [7] and [8] for infinite dimensional systems. All these authors found that they needed extra conditions on their class of systems in order to make the nice results of the finite dimensional case go through.

As is clear from the title that this monograph is about the geometric theory for infinite dimensional systems. In the next section we give an introduction to this problem similar to the finite dimensional case.

Disturbance Decoupling Problem for Infinite Dimensional Systems.

The class of systems that we shall consider in this section is the class of systems that have the representation

$$\dot{x}(t) = Ax(t) + Bu(t) + Eq(t);$$

(14) $\qquad y(t) = Cx(t);$

$$z(t) = Dx(t); \quad t \geq 0; \quad x(0) = x_0$$

where

$x(.), \quad u(.), \quad q(.), \quad y(.)$ and $z(.)$ are time trajectories in respectively $\mathcal{X}, \mathbf{R}^m, \mathcal{Q}, \mathcal{Y}$ and $\mathcal{Z}$, where $\mathcal{X}, \mathcal{Q}, \mathcal{Y}$ and $\mathcal{Z}$ are general Banach spaces,

B, C, D and E are bounded linear operators,

A is a generator of a C_0–semigroup $T_A(t)$.

These systems are called infinite dimensional after the infinite dimensionality of the state space $\mathcal{X}$. The notion of C_0–semigroup is the generalization of e^{At} from the finite dimensional case. Namely we have that $x(t) := T_A(t)x_0$ is the continuous solution of $\dot{x}(t) = Ax(t); \quad x(0) = x_0$. Furthermore the solution of (14) is given by, see Curtain and Pritchard [9, p. 31]

$$x(t) = T_A(t)x_0 + \int_0^t T_A(t-s)Bu(s)ds + \int_0^t T_A(t-s)Eq(s)ds;$$

(15) $\qquad y(t) = Cx(t) = CT_A(t)x_0 + \int_0^t CT_A(t-s)Bu(s)ds + \int_0^t CT_A(t-s)Eq(s)ds;$

$$z(t) = Dx(t) = DT_A(t)x_0 + \int_0^t DT_A(t-s)Bu(s)ds + \int_0^t DT_A(t-s)Eq(s)ds;$$

As in the finite dimensional part of this introduction we assume that we measure the full state of the system, i.e. we assume that $y(t) = x(t)$, or $C = I$. Furthermore we want again that the system Σ_f is as simple as possible, we therefore assume it to be time invariant and memory less. So we assume that

(16) Σ_f: $u(t) = Fx(t)$

for some F. In the finite dimensional case this feedback can always be respresented by a matrix. In infinite dimensional systems however there exists many different types of feedbacks that are linear and memory less. For mathematical convenience we shall make the extra assumption that F a **bounded** operator from the state space to $\mathbf{R}^m$. The problem at hand is to find a feedback system (16) such that after interconnecting (15) and (16) we have that the disturbance input $q(.)$ has no influence on the output $z(.)$. This problem we shall again define as the disturbance decoupling problem. From Curtain and Pritchard [9, p. 31] we have that the solution of the closed loop system is given by a formula similar to (5)

(17) $$z(t) = DT_{A+BF}(t)x_0 + \int_0^t DT_{A+BF}(t-s)Eq(s)ds;$$

So the DDP is solvable if and only if there exists a bounded feedback operator such that $DT_{A+BF}(.)E \equiv 0$. We can now apply a similar argument as in the finite dimensional case to link the DDP with invariant subspaces of the state space. For we have that DDP is solvable if and only if there exists a **closed** subspace V of the state space $\mathcal{X}$ with the properties

(18.a) $Im\ E \subset V \subset Ker\ D,$

(18.b) $T_{A+BF}(t)V \subset V;\ t \geq 0$

Once again we have that solving the disturbance decoupling problem is equivalent to finding subspaces that satisfy (18), only now we have the extra condition that V must be a closed subspace. In the finite dimensional case every subspace is closed. Thus it would be nice to have in our class a theorem similar to that on page 6. However since A and $A+BF$ are unbounded operators with the same domain we have that Av is not well defined for a general element $v \in \mathcal{X}$, but only for $v \in D(A)$, the domain of A. So the notions $ii)$ and $iii)$ in the theorem on page 6 must be modified to *ii)* $(A+BF)(V\cap D(A))\subset V$ for some bounded F.

iii) $A(V\cap D(A))\subset V+Im\,B$

We can pose the question whether these two concepts are equivalent, and whether one of them is equivalent to

i) $T_{A+BF}(t)V\subset V;\ t\geq 0$ for some bounded F

It is easy to prove that for closed subspaces *i)* implies *ii)* and *ii)* implies *iii)*. Unfortunately the converse does not hold in general. This has been proved by Pandolfi [27] for the implication *ii)* to *iii)*, and by Schmidt and Stern [32] for the implication *ii)* to *i)*.

So already in the beginning of the geometric theory for infinite dimensional systems there are negative results. The loss of equivalence between *i)* and *iii)* is very unfortunate, since *iii)* is a simple concept to work with. This combined with the equivalence between *i)* and *iii)* in the finite dimensional case, has made condition *iii)* into the key concept for the geometric theory in finite dimensional systems. So if we want to develop a geometric theory we have to develop concepts other than (A,B)–invariance.

As in the finite dimensional case we can pose the question if there exists a supremal subspace in the kernel of D, which satisfies (18.b). This subspace will again be denoted by $V^*(Ker\,D)$. The use of such a subspace will become clear from the next theorem by Curtain [6].

Theorem

If $V^*(Ker\,D)$ exists, then DDP is solvable if and only if $ImE\subset V^*(Ker\,D)$.

Furthermore the feedback that solves the DDP can be any bounded feedback F that satisfies $T_{A+BF}(t)V^*(Ker\,D)\subset V^*(Ker\,D)$.

Here again we have the negative result that $V^*(Ker\,D)$ does not necessarily exist, as was shown by Pandolfi [27] and Zwart [47], [48].

In this monograph we shall try to gain insight into these negative results. We introduce new concepts of invariance and we investigate whether they are equivalent to *i)*. In particular the follow new concept of invariance proves very important.

iv)   For every $x_0\in V$ there exists a continuous input $u(t)$ such that the solution of $\dot{x}(t)=Ax(t)+Bu(t);\ x(0)=x_0$ remains in V for all $t\geq 0$.

Outline of this Monograph

This work consists of 6 chapters and one appendix. The first chapter will be of an introductory nature and contains results on the relation between invariant subspaces of the system operator A and of its semigroup $T_A(t)$. In chapter II the main result of this monograph will be proved, that is the equivalence between the concepts i) and iv) for closed subspaces V. Furthermore we shall prove that these systems invariance concepts are equivalent to the concept of frequency invariance, which may be regarded, by Laplace transformation, as the frequency domain version of iv). This concept was introduced for finite dimensional systems by Hautus [19] and the generalization of it to the infinite dimensional case constituted a new starting point for the geometric theory for this class of systems. In chapter III the concept of frequency invariance is used to investigate the solvability of the DDP. The relation between the zeros of a single input system and its system invariant subspaces is used in chapter IV to give a complete characterization of all closed subspaces contained in the kernel of D and satisfying i). It turns out that the problem of the existence of $V^*(Ker\, D)$ is equivalent to the solvability of a pole placement problem. In chapter V we shall investigate disturbance decoupling for the case that we do not measure the whole state, i.e. $C \neq I$. Chapter VI is analogous, but also considering stability. In the appendix E we shall present counterexamples to certain properties that hold in the finite dimensional case, but not longer in the infinite dimensional case.

CHAPTER I: INVARIANCE CONCEPTS

In this chapter we shall consider the uncontrolled system

(1.1) $$\dot{x}(t) = Ax(t); \ x(0) = x_0$$

where A is the infinitesimal generator of the C_0–semigroup $T_A(t)$ on a Banach space X and x_0 is an arbitrary element of X. For standard theory concerning these generators we refer the reader to Curtain & Pritchard [9, ch. 2]. In this chapter we discuss some concepts of invariance for the system (1.1).

Section I.1: A– and $T_A(t)$-Invariance.

In this section we shall investigate invariance concepts related to the system (1.1). For the system (1.1) we shall define two concepts of invariance: semigroup invariance and generator invariance. If the state space X is finite dimensional, then these two concepts are equivalent. However, if X is infinite dimensional (this is the case we are mainly interested in), then they are in general not equivalent for unbounded generators of C_0–semigroups, example I.6. We shall first define semigroup– and generator invariance.

Definition I.1: Semigroup Invariance or $T_A(t)$-Invariance
A closed linear subspace V of X will be called semigroup invariant or $T_A(t)$-invariant if $T_A(t)V \subset V$, for all $t \geq 0$.

Definition I.2: Generator Invariance or A-Invariance
A closed linear subspace V of X will be called generator or A-invariant if $A(V \cap D(A)) \subset V$ where $D(A)$ is the domain of A.

So $T_A(t)$ invariance tells you that if we start in V, then the solution of (1.1) remains in V for all $t \geq 0$.

We shall recall some basic facts about these definitions. One of the most important questions is of course if these two concepts of invariance are equivalent. The next lemma will give a partial answer to this problem.

Lemma I.3.

Let A be the generator of the semigroup $T_A(t)$.

a) If a closed subspace is semigroup invariant, then it is also generator invariant.

b) If A is a bounded operator on $\mathcal{X}$, then semigroup and generator invariance are equivalent for closed subspaces.

Proof:

See Schmidt and Stern [32].

□

So the question remains whether or not semigroup and generator invariance are in general equivalent. The following lemma is very useful in order to answer this question.

Define ρ_∞ as the largest connected subset of $\rho(A)$ that contains an interval of the form $[r, +\infty)$. Since A generates a C_0-semigroup this set is nonempty, see Curtain and Pritchard [9, p.17].

Lemma I.4.

Let V be a closed linear subspace of $\mathcal{X}$, then the following concepts are equivalent

a) V is semigroup invariant.

b) $(\lambda I - A)^{-1}V \subset V$ for a λ in ρ_∞.

c) $(\lambda I - A)^{-1}V \subset V$ for all λ in ρ_∞.

d) The range of $(\lambda I - A)$ restricted to V is V for all $\lambda \in \rho_\infty$.

Proof:

a) → *b)*:see Pazy [29, p.121].

b) → *c)*:see Kurtz [23].

c) → *a)*:see Pazy [29, p.121].

a) ↔ *d)*:see Kurtz [23].

□

We remark here that the lemma is in general false (even for bounded generators) if ρ_∞ in *b)*, *c)* or *d)* is replaced by $\rho(A)$, as can be seen from the next example.

Example I.5.

This example will show that $T_A(t)$–invariance is in general <u>not</u> equivalent to $(\lambda - A)^{-1}$–invariance for all $\lambda \in \rho(A)$.

Let $\mathcal{X}$ be $\ell^\infty(\mathbb{Z})$, the space consisting of all $x:\mathbb{Z} \mapsto \mathbb{C}$ such that $\sup_{i \in \mathbb{Z}} |x(i)| < \infty$, and let A be the right shift on $\mathcal{X}$. So if e_i is the i–th basis vector, $e_i(j) = \delta_{ij}$, then $Ae_i = e_{i+1}$. This is a bounded operator and the spectrum of A is equal to the unit circle. If we define the set $V := \overline{\text{span}}\{e_i; \ i \in \mathbb{N}\}$, then V is obviously A–invariant. Since A is a bounded operator we have from lemma I.3 that V is $T_A(t)$–invariant. However A^{-1} is the left shift and since $A^{-1}e_0 = e_{-1}$, V is not A^{-1} invariant. □

For closed subspaces we have that semigroup invariance always implies generator invariance, but the converse is in general not true for unbounded generators, as the next example will show.

Example I.6.

Let $\mathcal{X}$ be the Hilbert space $L^2([0,1])$; the space of all functions $f(.)$ from $[0,1]$ to $\mathbb{R}$ such that $\int_0^1 |f(s)|^2 ds < \infty$. Furthermore let A be the "heat operator", the second derivative on $\mathcal{X}$ with domain equal to all functions $f(s)$ that are zero for $s = 0$ and $s = 1$ and such that the second derivative is in $\mathcal{X}$. This operator has a discrete spectrum, so $\rho_\infty = \rho(A)$. We define V to be the following subspace:

$$V = \{f \in L^2([0,1])| \ f(s) = 0 \ almost \ everywhere \ on \ [0, \tfrac{1}{2}]\}$$

Then V is A–invariant. However, if $\mathbb{1}_{[\frac{1}{2},1]}(s)$ denotes the indicator function of the interval $[\frac{1}{2},1]$, then

$$A^{-1}\left[\mathbb{1}_{[\frac{1}{2},1]}(s)\pi^2 sin(\pi s)\right] = \begin{cases} -s & ; \ s \in [0, \tfrac{1}{2}] \\ -s + 1 - sin(\pi s); & s \in [\tfrac{1}{2}, 1] \end{cases}$$

is not in V. Now lemma I.4 gives the desired result. □

So we have that for unbounded generators A– and $T_A(t)$–invariance is not in general equivalent. However, there is a class of subspaces where the two concepts of invariance are equivalent.

Lemma I.7.

If V is a closed linear subspace contained in the domain of A, then generator invariance is equivalent to semigroup invariance.

Proof:

see Curtain [6, lemma 2.3.].

□

The next lemma will give an important property of $T_A(t)$–invariant subspaces.

Lemma I.8.

Let V be a closed linear subspace which is $T_A(t)$-invariant, then $\overline{V \cap D(A)} = V$.

Proof:

From page 37 of Davies [14] we have that every element x in $\mathcal{X}$ is the limit of $\lambda(\lambda I - A)^{-1}x$, as $\lambda \to +\infty$.

Assume that $x \in V$, then by lemma I.4. we have, for λ sufficiently large, that $(\lambda I - A)^{-1}x$ is in $V \cap D(A)$.

Combining these results we have that x is the limit of a sequence in $V \cap D(A)$. So $\overline{V \cap D(A)} = V$.

□

As can be seen from example I.6 we have that A–invariance is weaker than $T_A(t)$–invariance even if we assume the extra condition that $V \cap D(A)$ is dense in V.

Section I.2: The Relation between $T_A(t)$-Invariance and the spectrum of A

In this section we shall investigate the relation between the spectrum of A, $\sigma(A)$, and the semigroup invariant subspaces. We remark that if V is a closed A–invariant subspace, then the restriction of A to V is a well defined closed operator on V. We shall denote this operator by A_V.

Again let ρ_∞ denote the largest connected subset of the resolvent set of A that contains an interval of the form $[r,\infty)$. Now we have the next lemma.

19

Lemma I.9.

Let A generate the C_0–semigroup $T_A(t)$ and let V be a closed linear subspace of X. Then the following assertions are equivalent:

i) V is $T_A(t)$–invariant.

ii) V is A–invariant and $\rho_\infty \subset \rho(A_V)$.

Proof:

  $i) \rightarrow ii)$

      This follows directly from lemma I.3.a) and I.4.d).

  $ii) \rightarrow i)$

      Let V be an A–invariant subspace with $\rho_\infty \subset \rho(A_V)$. Then for all λ in $\rho(A_V)$ we have that the range of $\lambda I_V - A_V$ is equal to V, but by the definition of A_V we have that this implies that the range of $\lambda - A$ restricted to V is equal to V. Lemma I.4 gives the desired result.   □

This result is the strongest one that we can obtain, since in example I.5 we have that $\rho_\infty = \rho(A_V) = \{s \in \mathbb{C} : |s| > 1\}$, see Kato [22, p. 210].

From this lemma we have the following corollary.

Corollary I.10.

Suppose that the resolvent set of A is connected and that V is a closed A–invariant subspace. Then V is $T_A(t)$–invariant if and only if $\sigma(A_V) \subset \sigma(A)$.

Proof:

This is a direct consequence of the above lemma. Since if $\rho(A)$ is connected, then $\rho(A) = \rho_\infty$.   □

So for example I.6 this corollary implies that there $\sigma(A_V) = \mathbb{C}$.

Chapter II: System Invariance Concepts

The theory of controlled invariance of a subspace has been investigated in detail in the case that the state space is finite dimensional, see e.g. [1], [19], [35] and [42]. For the case that the state space is infinite dimensional there have been some preliminary investigations in [6]–[8], [27] and [32], but many questions remain unanswered. In this chapter we shall investigate this invariance for infinite dimensional systems for the case of finite rank inputs. We shall consider the controlled version of system (1.1):

$$(2.1) \qquad \dot{x}(t) = Ax + Bu; \ x(0) = x_0, \ x \in \mathcal{X}, \ u \in \mathcal{U},$$

where $\mathcal{X}$ and $\mathcal{U}$ are Banach spaces, A is a generator of a C_0–semigroup, $T_A(t)$, and furthermore we shall impose the condition that B is a bounded linear operator with $Im \, B$ finite dimensional, so without loss of generality we may assume that $\mathcal{U} = \mathbb{R}^m$ and B is injective. The reason for only considering finite rank inputs operators is twofold. First of all it is from a practical point of view a natural choice, one can only implement finitely many inputs. Secondly there is the mathematical reason; we believe that only a small part of the theory, as will be presented here, will remain valid if $Im \, B$ is not finite dimensional.

For the system (2.1) we shall discuss various kinds of system invariance in section II.1, and we shall pay particular attention to open loop invariance (section II.2.) and frequency invariance (section II.3.). The concept of frequency invariance was introduced by Hautus [19] for finite dimensions, and it turns that this concept plays a key role in infinite dimensions. In section II.4. the relation between the various invariance concepts is investigated, and there we show that for closed subspaces, open and closed loop invariance are equivalent. Furthermore we prove that for these subspaces closed loop invariance is equivalent to frequency invariance.

Section II.1: System Invariance Concepts

The theory of system invariance entails many definitions. Here we shall summarize some of them and give some important properties. We shall start with the strongest.

By $T_{A+BF}(t)$ we shall denote the semigroup generated by $A+BF$. Since F is a bounded operator we have from Curtain and Pritchard [9, p. 38] that $A+BF$ always generates a C_0–semigroup and the domain of $A+BF$ is equal to the domain of A.

Definition II.1: Closed Loop Invariance

A subspace V of $\mathcal{X}$ is called closed loop invariant if there exists a bounded feedback law F such that

$$(2.2) \qquad T_{A+BF}(t)V \subset V$$

for all t in $[0,\infty)$.

Remark:

So a subspace V is called closed loop invariant if it is semigroup invariant for the system $\dot{x}(t) = (A+BF)x(t)$ for some $F \in \mathcal{L}(\mathcal{X},\mathcal{U})$.

Lemma II.2.

Assume that a closed linear subspace $V \subset \mathcal{X}$ is $T_{A+BF_1}(t)$–invariant, for a certain bounded operator F_1. Then V is $T_{A+BF_2}(t)$–invariant for a bounded operator F_2 if and only if $Im\, B(F_1 - F_2)|_{V \cap D(A)} \subset V$.

Proof:

See lemma 4 in Curtain [7]. □

Corollary II.3.

Let $\hat{B}$ be any subspace of $Im\, B$ such that $\hat{B} + (Im\, B \cap V) = Im\, B$.

If a closed subspace V is closed loop invariant, then there exists a bounded feedback law F such that V is $T_{A+BF}(t)$–invariant and $Im\, BF\Big|_{V \cap D(A)} \subset \hat{B}$.

Proof:

Assume that V is $T_{A+B\tilde{F}}(t)$–invariant, then from the fact that the range of $\tilde{F}$ is finite–dimensional and $\tilde{F}$ is bounded, $B\tilde{F}$ can be written as,

(Kato [22, p.160]),

$$\sum_{i=1}^{q} b_i <.,f_i> + \sum_{i=q+1}^{m} b_i <.,f_i>, \text{ where } \operatorname*{span}_{i=1,.,q} \{b_i\} = \hat{B} \text{ and } \operatorname*{span}_{i=q+1,.,m} \{b_i\} \subset \operatorname{Im} B \cap V.$$

Defining $F = \left[<.,f_i> \right]_{i=1}^{q}$ we obtain that $\operatorname{Im} B(F-\tilde{F})|_{V\cap D(A)} \subset V$ and $BF|_{V\cap D(A)} \subset \hat{B}$.

Thus lemma II.2 concludes the assertion.

$\square$

Now we shall define the generator invariance corresponding to closed loop invariance. In the sequel of this chapter it will turn out that it is convenient to use a larger class of feedback operators in this definition. We shall use the class of A-bounded operators.

Definition II.4: A-Bounded Operator

Let λ be an element of $\rho(A)$.
An operator F from $\mathcal{X}$ to $\mathcal{U}$ will be A-bounded if $D(A) \subset D(F)$ and $F(\lambda-A)^{-1}$ is a bounded operator from $\mathcal{X}$ to $\mathcal{U}$.

It is easy to see that if F is a bounded operator, then it is also an A-bounded operator. So the class of A-bounded operators is larger than the class of bounded operators. If A is a bounded operator itself, then these classes are equal. Furthermore, if $F(\lambda_0-A)^{-1}$ is a bounded operator from $\mathcal{X}$ to $\mathcal{U}$, then by the resolvent identity $F(\lambda-A)^{-1}$ is a bounded operator from $\mathcal{X}$ to $\mathcal{U}$ for all $\lambda \in \rho(A)$. So definition II.4 is independent of the particular choice of $\lambda \in \rho(A)$

Definition II.5: Feedback Invariance

A subspace V of $\mathcal{X}$ is called feedback invariant if there exists an A-bounded feedback law F such that

$$(2.3) \qquad (A+BF)(V\cap D(A)) \subset V$$

Definition II.6: (A,B)-Invariance

A subspace V of $\mathcal{X}$ is called (A,B) invariant if

$$(2.4) \qquad A(V\cap D(A)) \subset V + \operatorname{Im} B.$$

Remark:

Let $\hat{B}$ be any subspace of $Im\,B$ such that $\hat{B}+(Im\,B\cap V)=Im\,B$. Then by a proof similar to that in corollary II.3, it can be shown that if a subspace is feedback invariant, it is also feedback invariant for an A-bounded feedback satisfying $Im\,BF|_{V\cap D(A)}\subset\hat{B}$, see Zwart [48].

In the next two sections we shall define two other invariance concepts which are strongly related with the solutions of equation (2.1). These two invariant concepts are closely related, i.e. the first one is the time domain version of invariance and the second is the frequency domain description of the same property. As is to expected these concepts have many properties in common.

Section II.2: Open Loop Invariance

Here we shall make precise what we mean if we say that the solution of (2.1) remains in V. Furthermore we shall give some interesting properties of this concept, that are useful in the sequel.

By the solution of (2.1) for $u(.):[0,\infty)\mapsto\mathcal{U}$ we mean the mild solution:

$$(2.5) \qquad x(t)=T_A(t)x_0+\int_0^t T_A(t-s)Bu(s)ds$$

Definition II.7: Open Loop Invariance

A subspace $V\subset\mathcal{X}$ is said to be open loop invariant if for every $x_0\in V$ there exists a <u>continuous</u> $u(.):[0,\infty)\mapsto\mathcal{U}$ such that the solution of (2.1) remains in V.

Remark:

If the state space is finite dimensional, then it is standard to allow arbitrary measurable functions in the definition of open loop invariance. As a result one obtains there that the input function could be chosen to be continuous. However if the state space is infinite dimensional, then this result no longer holds, see Zwart [48]. So we have to restrict the class of input functions in this definition and here we have restricted ourselves to continuous functions. A result that is still unproved, but is believed to hold, is that we could also use piecewise continuous functions. Then our definition of open loop invariance would be the same as the one used by J.v.d. Woude [43].

Example II.8.

In this example we shall study the delay system

(2.6) $\dot{y}(t) = y(t-1) + u(t); \ t \geq 0 \ and \ y(t) = y_0(t); \ -1 \leq t \leq 0$

This system can be rewritten in the form (2.1) if we make the following definitions.

$$M^2((-1,0);R) = R \oplus L^2((-1,0);R)$$
$$A:D(A) \mapsto M^2((-1,0);R) \ with$$
$$D(A) = \{(\varphi_0, \varphi) \in M^2((-1,0);R) | \varphi_0 \in R, \varphi \ such \ that \ \varphi' \in L^2((-1,0);R)\}$$
$$A(\varphi_0, \varphi) = (\varphi(-1), \varphi')$$
$$B:R \mapsto M^2((-1,0);R) \ with$$
$$Bu = (u, 0).$$

It can be shown that A generates a semigroup on $M^2((-1,0);R)$, see Curtain & Pritchard [9, p.48].

We shall show that the subspace

$$V: = \{0\} \oplus \{\varphi \in L^2((-1,0);R)| \ \varphi \ is \ continuous \ on \ [-1,0] \ and \ \varphi(0) = 0\}$$

is open loop invariant.

Let $(0, \varphi)$ be an element of V, then by defining $u(t) = -\varphi(t-1)$ for $0 \leq t \leq 1$ we have that

$$\dot{y}(t) = y(t-1) + u(t) = 0 \ for \ 0 \leq t \leq 1 \ and \ y(0) = 0$$

This implies that $y(t) = 0$ for $0 \leq t \leq 1$. Now we define $u(t)$ to be zero for $t \geq 1$, and with (2.6) this gives that $y(t) \equiv 0$ for $t \geq 0$. The state at time t is $(y(t), \ y(t-\tau); \ 0 \leq \tau \leq 1)$ and this is in V for all $t \geq 0$. Furthermore by the continuity of $y(.)$ for $t \geq -1$ we have that $u(.)$ is continuous. So V is open loop invariant.

$\square$

In the sequel of this section we shall prove some technical results concerning open loop invariance. We advise the reader who is not interested in details to proceed to section II.3.

In the proof of the properties of open loop invariance it is very useful to have some results concerning the convolution in (2.5). This will be given in the next technical lemma.

Lemma II.9.

Let $*$ denote the convolution product, and let $k(.)$ be a continuous function with values in $\mathbb{R}^{n*n}$ (the n by n matrices). Then there exists a continuous function $k^{inv}(.)$ such that if $f(.)$, $g(.) \in C([0,\infty);\mathbb{R}^n)$ satisfy the following convolution equation

(2.7)
$$g(t) = f(t) + k(t)*f(t),$$

then they satisfy also

$$f(t) = g(t) - k^{inv}(t)*g(t),$$

Proof:

See Doetsch [16, ch. 40].   □

Let V be a subspace of $\mathcal{X}$. By $B^0(V)$ and $B^1(V)$ we shall denote subspaces of $\mathcal{X}$ such that

(2.8)
$$\begin{cases} Im\, B = B^0(V) \oplus B^1(V), \\ B^0(V) \cap V = \{0\} \text{ and} \\ B^1(V) = V \cap Im\, B \end{cases}$$

We remark that (2.8) is possible, since $Im\, B$ is finite dimensional. So $B^0(V)$ is a subspace contained in $Im\, B$ which has zero intersection with V and is of maximal dimension under this restriction. Throughout this section we shall fix $B^0(V)$. So if $\mathcal{X}$ is a Hilbert space, then we can take $B^0(V) = Im\, B \cap (Im\, B \cap V)^{\perp}$. If there cannot be any doubt about V, then we shall simply use B^0 and B^1.

With this notation we shall prove that if a closed subspace is open loop invariant for the system (A,B), then it is also open loop invariant for the system (A,B^0), where $Im\, B^0 = B^0$.

By $C([0,\infty),\mathcal{U})$ we shall denote the space of all continuous functions from $[0,\infty)$ to $\mathcal{U}$. We stress that we do not impose any norm on this space.

Lemma II.10.

If V is a closed subspace of $\mathcal{X}$ that is open loop invariant, then for every x_0 in V there exists *one and only one* $u(.)\in C([0,\infty);\mathcal{U})$ such that the solution of (2.1) remains in V and $Bu(.)\in B^0$.

Remark:

This result tells us that control actions with values in B^1 can be omitted. Or from a different perspective; the only freedom we have in choosing our control action is in B^1. So if $u_1(.)$ and $u_2(.)$ are both controls that keep x_0 in V, then $B\{u_1(.)-u_2(.)\}\in B^1$.

This result is intuitively not hard to understand. Since if we start in zero at $t=0$ and the control action $Bu(.)$ is in V, then since $\dot{x}(0)=Bu(0)$ we shall remain in V. Otherwise if the control action is not in V, then we shall leave V.

Proof:

Let $\{b_1,..,b_{m_0}\}$ be a basis for B^0 and let $\{b_{m_0+1},..,b_m\}$ be a basis for B^1. We shall begin with showing that for every b_j; $m_0+1\leq j\leq m$, there exists an input $\hat{u}_j(.)$ such that $B\hat{u}_j(.)\in B^0$ and if we start in b_j $(x_0=b_j)$, then the solution of (2.1) with this $\hat{u}_j(.)$ remains in V.

Since V is open loop invariant we have for every b_j in B^1 the existence of a $u_j(.)=[u_{1,j},..,u_{m,j}]^T\in C([0,\infty);\mathcal{U})$ such that if one starts in b_j at time instant zero, then one remains in V. Now rewriting this solution of (2.1) gives

$$x_j(t)=T_A(t)b_j + \sum_{i=1}^{m_0}\int_0^t T_A(t-s)b_iu_{i,j}(s)ds + \sum_{i=m_0+1}^{m}\int_0^t T_A(t-s)b_iu_{i,j}(s)ds,$$

is in V for $m_0+1\leq j\leq m$.

Let $*$ denote the convolution product, then this equation can be rewritten in the following form

$$x_j(t)=T_A(t)b_j + [u_{1,j}(.),..,u_{m_0,j}(.)]*\begin{bmatrix}T_A(.)b_1\\ \vdots\\ T_A(.)b_{m_0}\end{bmatrix} +$$

$$[u_{m_0+1,j}(.),..,u_{m,j}(.)]*\begin{bmatrix}T_A(.)b_{m_0+1}\\ \vdots\\ T_A(.)b_m\end{bmatrix}; \quad m_0+1\leq j\leq m$$

Writing this in matrix notation for $j = m_0+1,..,m$, we have

$$\begin{bmatrix} x_{m_0+1}(t) \\ \vdots \\ x_m(t) \end{bmatrix} = \begin{bmatrix} T_A(t)b_{m_0+1} \\ \vdots \\ T_A(t)b_m \end{bmatrix} + \begin{bmatrix} u_{1,m_0+1}(\cdot) & \cdots & u_{m_0,m_0+1}(\cdot) \\ \vdots & & \vdots \\ u_{1,m}(\cdot) & \cdots & u_{m_0,m}(\cdot) \end{bmatrix} * \begin{bmatrix} T_A(\cdot)b_1 \\ \vdots \\ T_A(\cdot)b_{m_0} \end{bmatrix} +$$

$$\begin{bmatrix} u_{m_0+1,m_0+1}(\cdot) & \cdots & u_{m,m_0+1}(\cdot) \\ \vdots & & \vdots \\ u_{m_0+1,m}(\cdot) & \cdots & u_{m,m}(\cdot) \end{bmatrix} * \begin{bmatrix} T_A(\cdot)b_{m_0+1} \\ \vdots \\ T_A(\cdot)b_m \end{bmatrix}$$

By introducing the following notation we can simplify the above equation.

Let $U_{m_0} := \begin{bmatrix} u_{1,m_0+1}(\cdot) .. u_{m_0,m_0+1}(\cdot) \\ \vdots \qquad \vdots \\ u_{1,m}(\cdot) .. u_{m_0,m}(\cdot) \end{bmatrix}$ and $U_{m-m_0} := \begin{bmatrix} u_{m_0+1,m_0+1}(\cdot) .. u_{m,m_0+1}(\cdot) \\ \vdots \qquad \vdots \\ u_{m_0+1,m}(\cdot) .. u_{m,m}(\cdot) \end{bmatrix}$.

So these matrices are named after their number of colomns.

Rewriting the above equation with this notation gives

$$\begin{bmatrix} x_{m_0+1}(t) \\ \vdots \\ x_m(t) \end{bmatrix} - U_{m_0}* \begin{bmatrix} T_A(\cdot)b_1 \\ \vdots \\ T_A(\cdot)b_{m_0} \end{bmatrix} = \begin{bmatrix} T_A(t)b_{m_0+1} \\ \vdots \\ T_A(t)b_m \end{bmatrix} + U_{m-m_0}* \begin{bmatrix} T_A(\cdot)b_{m_0+1} \\ \vdots \\ T_A(\cdot)b_m \end{bmatrix}$$

From lemma II.9 we have that there exists a function $U_{m-m_0}^{inv}(t)$ in $C([0,\infty);\mathbb{R}^{m-m_0,m-m_0})$ such that

$$(2.9) \qquad \begin{bmatrix} T_A(t)b_{m_0+1} \\ \vdots \\ T_A(t)b_m \end{bmatrix} = \begin{bmatrix} x_{m_0+1}(t) \\ \vdots \\ x_m(t) \end{bmatrix} - U_{m_0}(\cdot)* \begin{bmatrix} T_A(\cdot)b_1 \\ \vdots \\ T_A(\cdot)b_{m_0} \end{bmatrix} -$$

$$U_{m-m_0}^{inv}(\cdot)* \left\{ \begin{bmatrix} x_{m_0+1}(t) \\ \vdots \\ x_m(t) \end{bmatrix} - U_{m_0}(\cdot)* \begin{bmatrix} T_A(\cdot)b_1 \\ \vdots \\ T_A(\cdot)b_{m_0} \end{bmatrix} \right\}.$$

Rearranging this equation gives

$$\begin{bmatrix} x_{m_0+1}(t) \\ \vdots \\ x_m(t) \end{bmatrix} - U_{m-m_0}^{inv}(t)* \begin{bmatrix} x_{m_0+1}(t) \\ \vdots \\ x_m(t) \end{bmatrix} = \begin{bmatrix} T_A(t)b_{m_0+1} \\ \vdots \\ T_A(t)b_m \end{bmatrix} +$$

$$\left\{ U_{m_0}(\cdot) - U_{m-m_0}^{inv}(\cdot)*U_{m_0}(\cdot) \right\}* \begin{bmatrix} T_A(\cdot)b_1 \\ \vdots \\ T_A(\cdot)b_{m_0} \end{bmatrix}.$$

For $j = m_0+1,..,m$ let $\hat{x}_j(\cdot)$ and $\hat{u}_j(\cdot)$ denote respectively $j-m_0$ th element of

$$\begin{bmatrix} x_{m_0+1}(t) \\ \vdots \\ x_m(t) \end{bmatrix} - U_{m-m_0}^{inv}(t)* \begin{bmatrix} x_{m_0+1}(t) \\ \vdots \\ x_m(t) \end{bmatrix}$$ and the transpose of the $j-m_0$ th row of

$U_{m_0} - U^{inv}_{m-m_0} * U_{m_0}$. Then the equation above gives

(2.10)
$$\hat{x}_j(.) = T_A(.)b_j + \hat{u}_j(.)^T * \begin{pmatrix} T_A(.)b_1 \\ \vdots \\ T_A(.)b_{m_0} \end{pmatrix}.$$

Since V is a closed subspace and $x_j(.) \in V$ we have that $\hat{x}_j(.) \in V$, $j = m_0 + 1, .. m$. Furthermore $U^{inv}_{m-m_0}(.) * U_{m_0}(.)$ is by the continuity of $U^{inv}_{m-m_0}(.)$ and $U_{m_0}(.)$ and by standard convolution theory an element of $C([0,\infty);R^{m-m_0,m_0})$ and so $\hat{u}_j(.)$ is an element of $C([0,\infty);R^{m_0})$. With this input function we can define an input function in $C([0,\infty);R^m)$ by just adding $m - m_0$ rows with the zero function. This extended input function we shall again denote by $\hat{u}(.)$. So from equation (2.10) we have that for every b_j in B^1 there exists an $\hat{u}_j(t) \in C([0,\infty);R^m)$ such that the solution of (2.1) remains in V and $B\hat{u}_j(t) \in B^0$ for all $t \geq 0$.

Thus we have proved our assertion for $x_0 \in B^1$. Let x_0 be an arbitrary element of V, then there exists an input $u(t)$ such that

$$x(t) = T_A(t)x_0 + \int_0^t T_A(t-s)Bu(s)ds$$

remains in V. Using the fact that $u(t) = [u_1(t), .., u_m(t)]^T$ we have that

$$x(t) = T_A(t)x_0 + \sum_{i=1}^{m_0} \int_0^t T_A(t-s)b_i u_i(s)ds + \sum_{i=m_0+1}^{m} \int_0^t T_A(t-s)b_i u_i(s)ds$$

Or in matrix notation with $*$ again denoting the convolution product

$$x(t) = T_A(t)x_0 + [u_1(.), .., u_{m_0}(.)] * \begin{pmatrix} T_A(.)b_1 \\ \vdots \\ T_A(.)b_{m_0} \end{pmatrix} +$$

$$[u_{m_0+1}(.), .., u_m(.)] * \begin{pmatrix} T_A(.)b_{m_0+1} \\ \vdots \\ T_A(.)b_m \end{pmatrix}$$

Equation (2.9) gives that

$$\begin{pmatrix} T_A(t)b_{m_0+1} \\ \vdots \\ T_A(t)b_m \end{pmatrix} = \begin{pmatrix} x_{m_0+1}(t) \\ \vdots \\ x_m(t) \end{pmatrix} - U^{inv}_{m-m_0}(t) * \begin{pmatrix} x_{m_0+1}(t) \\ \vdots \\ x_m(t) \end{pmatrix} -$$

$$\left\{ U_{m_0}(.) - U^{inv}_{m-m_0}(.) * U_{m_0}(.) \right\} * \begin{pmatrix} T_A(.)b_1 \\ \vdots \\ T_A(.)b_{m_0} \end{pmatrix}.$$

Defining $\tilde{x}(.)$ as $[u_{m_0+1}(.),..,u_m(.)]*\left\{\begin{pmatrix} x_{m_0+1}(t) \\ \vdots \\ x_m(t) \end{pmatrix} - U^{inv}_{m-m_0}(t)*\begin{pmatrix} x_{m_0+1}(t) \\ \vdots \\ x_m(t) \end{pmatrix}\right\}$ and

$\tilde{u}(.) = [\tilde{u}_1(.),..,\tilde{u}_{m_0}(.)]^T$ as $[u_{m_0+1}(.),..,u_m(.)]*\left\{U_{m_0}(.) - U^{inv}_{m-m_0}(.)*U_{m_0}(.)\right\}$

gives

(2.11) $\qquad x(t) = T_A(t)x_0 + \left\{[u_1(.),..,u_{m_0}(.)] - \tilde{u}(.,)^T\right\}*\begin{pmatrix} T_A(.)b_1 \\ \vdots \\ T_A(.)b_{m_0} \end{pmatrix} + \tilde{x}(t)$

By the continuity of $u(.)$, $U_{m_0}(.)$ and $U^{inv}_{m-m_0}(.)$ plus standard convolution theory we have that $\tilde{u}_i(.) \in C([0,\infty);\mathbb{R})$ for $1 \leq i \leq m_0$. Furthermore since V is a closed subspace and $x_j(.) \in V; \ j = m_0+1,..,m$ we have that $\tilde{x}(.) \in V$.

$x(t) - \tilde{x}(t) = T_A(t)x_0 + [u_1(.) - \tilde{u}_1(.),..,u_{m_0}(.) - \tilde{u}_{m_0}(.)]*\begin{pmatrix} T_A(.)b_1 \\ \vdots \\ T_A(.)b_{m_0} \end{pmatrix}$

We have that $x(t)-\tilde{x}(t)$ is an element of V and $u_i(.) - \tilde{u}_i(.); \ 1 \leq i \leq m_0$ is a continuous function. If we define $\check{u}_i(.) = u_i(.) - \tilde{u}_i(.)$ for $1 \leq i \leq m_0$ and $\check{u}_i(.) = 0$ for $m_0+1 \leq i \leq m$, then $\check{u}(.) = [\check{u}_1(.),..,\check{u}_m(.)] \in C([0,\infty);\mathbb{R}^m)$, $B\check{u}(.) \in B^0$ and

$x(t) - \tilde{x}(t) = T_A(t)x_0 + \sum_{i=1}^{m_0} \int_0^t T_A(t-s)b_i\check{u}_i(s)ds + \sum_{i=m_0+1}^{m} \int_0^t T_A(t-s)b_i\check{u}_i(s)ds$

So we have proved that for every x_0 in V there exists an $\check{u}(t) \in C([0,\infty);\mathcal{U})$ such that $B\check{u}(t) \in B^0$.

It remains to prove that this input is unique. By the linearity of the system it is sufficient to prove that the input which holds $x_0=0$ in V is identically zero.

So suppose that $x(t) = \int_0^t T_A(t-s)Bu(s)ds$ is in V and $Bu(.) \in B^0$. Let $\{b_1,..,b_{m_0}\}$ be a basis for B^0. Then by the Hahn–Banach theorem there exist functionals $g_j \in \mathcal{X}'$ such that $<g_j, b_i> = \delta_{ij}$ and $g_j|_V = 0$. Operating these g_j on $x(t)$ gives

$\begin{pmatrix} 0 \\ \vdots \\ 0 \end{pmatrix} = \begin{pmatrix} <g_1,T_A(.)b_1> & .. & <g_1,T_A(.)b_{m_0}> \\ \vdots & & \vdots \\ <g_{m_0},T_A(.)b_1> & .. & <g_{m_0},T_A(.)b_{m_0}> \end{pmatrix} * \begin{pmatrix} u_1(.) \\ \vdots \\ u_{m_0}(.) \end{pmatrix}$

Since $\begin{bmatrix} <g_1,T_A(.)b_1> & \cdots & <g_1,T_A(.)b_{m_0}> \\ \vdots & & \vdots \\ <g_{m_0},T_A(.)b_1> & \cdots & <g_{m_0},T_A(.)b_{m_0}> \end{bmatrix}$ is in $t=0$ equal to the identity on R^{m_0} we have from standard convolution theory that $u(t)$ is identically zero.

$\square$

Lemma II.11.

Let V be a closed subspace that is open loop invariant and suppose that $B^1(V) = \{0\}$. Then for every x_0 in V, there exists a $u(.) \in C([0,\infty);\mathcal{U})$ which is exponentially bounded and such that the solution of (2.1) remains in V. By lemma we already have that this input is unique.

Proof:

From lemma II.10 we have that for all $x_0 \in V$ there exists an unique $u(.) \in C([0,\infty);\mathcal{U})$ such that the solution of (2.1) remains in V. Define Q to be the operator that assigns to x_0 in V this input restricted to $[0,1]$. The linearity of this operator follows obviously from the uniqueness of the input. If we denote by $C_{[0,1]}$ the Banach space of all continuous functions from $[0,1]$ to $\mathcal{U}$ with as norm the sup-norm on $[0,1]$, then, as we shall prove, Q is linear bounded operator from the Banach space V to $C_{[0,1]}$. Since Q is defined on the whole of V it is sufficient to prove that Q is a closed operator. Let x_n converge to x in V and let $u_n = Qx_n$ converge in $C_{[0,1]}$ to $u(.)$. By $x_n(t)$ we shall denote the solution of (2.1) for x_n and $u_n(.)$. So

$$x_n(t) = T_A(t)x_n + \int_0^t T_A(t-s)Bu_n(s)ds$$

Then we have that for all $t \in [0,1]$ the pointwise limit, $x(t)$, of $x_n(t)$ exists and satisfies $x(t) = T_A(t)x + \int_0^t T_A(t-s)Bu(s)ds$. Furthermore since V is a closed subspace $x(t)$ is in V. So $Qx=u$, and Q is closed, thus bounded.

Now let x_0 be an arbitrary element of V and let $u(.)$ be the input that keeps the solution $x(.)$ of (2.1) in V. So

(2.12) $$x(t) = T_A(t)x_0 + \int_0^t T_A(t-s)Bu(s)ds$$

We shall prove that this input $u(.)$ is exponentially bounded. Let x_i be $x(i)$, $q = \|Q\|$ and suppose that $\|T_A(t)\| \le Me^{\omega t}$, then

(2.13) $\quad \|x(1)\| \leq \|T_A(1)x_0 + \displaystyle\int_0^1 T_A(1-s)Bu(s)ds\| \leq Me^\omega\|x_0\| + Me^\omega\|B\|\left(\sup_{0\leq s\leq 1}\|u(s)\|\right)$

$$\leq Me^\omega\{1 + \|B\|q\}\|x_0\|. \text{ Since } \sup_{0\leq s\leq 1}\|u(s)\| = \|Qx\| \leq q\|x\|.$$

Let k be $Me^\omega\{1 + \|B\|q\}$, then we shall show that $\sup_{i\leq s\leq i+1}\|u(s)\| \leq q\|x_i\|$ and $\|x_{i+1}\| \leq k\|x_i\|$. By the time invariance of the system we have that for t larger than 0:

$$x(t+i) = T_A(t)x_i + \int_0^t T_A(t-s)Bu(i+s)ds$$

So $Qx_i = u(i+.)$ and thus $\sup_{i\leq s\leq i+1}\|u(s)\| = \sup_{0\leq s\leq 1}\|u(i+s)\| = \|Qx_i\| \leq q\|x_i\|$ and with a similar argument as in (2.13) we have that $\|x_{i+1}\| \leq k\|x_i\|$. So on the interval $[i, i+1]$ we have that $\|u(t)\| \leq q\|x_i\| \leq qk\|x_{i-1}\| \leq qk^i\|x_0\| \leq qe^{\{log(k)\}t}\|x_0\|$ and this proves that $u(.)$ is exponentially bounded. $\qquad\square$

With the above lemma one can easily prove that if a closed subspace is open loop invariant, then it is also closed loop invariant. Here we shall only give a sketch of this proof since in section II.4 we shall present a different proof. With the same notation as in the proof of II.11 we define the feedback law F as the composition of the bounded operator Q and the bounded operator $\Delta: C_{[0,1]} \mapsto \mathcal{U}$ given by $\Delta u = u(0)$. Then this F is a bounded operator and satisfies $Fx(0) = u(0)$. With the time invariance of the system we have $Fx(t) = u(t)$, where $u(.)$ is the unique continuous input that holds $x(.)$ in V.

So the above lemma implies that the $u(.)$ in lemma II.10 is exponentially bounded. In this lemma we have restricted our attention to closed subspaces. In section III.3 we shall give a sufficient condition such that if V is open loop invariant, then the same holds for the closure of V.

Section II.3: Frequency Invariance

In this section we shall give a frequency domain version of open loop invariance, this concept will be called frequency invariance. In Hautus [19] the concept of frequency invariance appeared for the first time. Only the finite dimensional case was discussed there. In this section we shall discuss this concept for infinite dimensional systems.

The next definition will generalize the concept of the space of all rational functions.

Definition II.12: $Y(s)$ and $Y_{-1}(s)$

Let Y be a subspace, not necessarily closed, of a Banach space Z. A function $f(.)$ is an element of $Y(s)$ if $f(.)$ is defined and continuous on some interval of the form $[r,\infty)$ and $f(s_0)$ is an element of Y for all s_0 in this interval.

By $Y_{-1}(s)$ we shall denote all functions in $Y(s)$ that are of the order s^{-1} for $s \to \infty$ i.e.

$$Y_{-1}(s) = \{f \in Y(s) \mid \lim_{s \to \infty} sf(s) \text{ exists}\}$$

These functions will be called *strictly proper*.

We remark that instead working with these definitions we could also follow Zwart [48] where $Y(s)$ was defined as the set of all functions which are meromorphic in some right half-plane and with values in Y, and $Y_{-1}(s)$ as those meromorphic functions with values in Y with $\lim_{\substack{s \to \infty \\ s \in \mathbb{R}}} sf(s)$ exists. We have chosen the definitions above, since these will shorten some proofs.

Definition II.13: (ξ,ω)-Representation

If $x_0 \in X$, $\xi(.) \in D(A)(s)$ and $\omega(.) \in U_{-1}(s)$, then the expression

$$(2.14) \qquad x_0 = (s-A)\xi(s) - B\omega(s)$$

is called a (ξ,ω)-representation of x_0.

We remark that (2.14) is only defined for $s \in [r, \infty)$, for some $r > 0$. Furthermore we have that every x_0 in $\mathcal{X}$ has a (ξ, ω)–representation: take $\xi(s) = (s - A)^{-1} x_0$ and $\omega(s) = 0$, for $s \in \rho(A) \cap \mathbb{R}$.

The (ξ, ω)–representation has the following important property.

Lemma II.14.

If $\xi(s)$, $\omega(s)$ is a (ξ, ω)–representation of x, then $\xi(s) \in D(A)_{-1}(s)$ and $\lim_{s \to \infty} s \xi(s) = x$.

Proof:

By the fact that A generates a C_0–semigroup we have that $\lim_{s \to \infty} s(s - A)^{-1} x = x$ for all x in $\mathcal{X}$. By the (ξ, ω)–representation

$$x = (s - A)\xi(s) - B\omega(s) \text{ or } s\xi(s) = s(s - A)^{-1} x - s(s - A)^{-1} B\omega(s).$$

Thus $\lim_{s \to \infty} s\xi(s) = \lim_{s \to \infty} s(s - A)^{-1} x - \lim_{s \to \infty} s(s - A)^{-1} B\omega(s) = x + 0$, since B is bounded and $\lim_{s \to \infty} \omega(s) = 0$. □

With this (ξ, ω)–representation we can introduce the concept of frequency invariance.

Definition II.15: Frequency Invariance.

Let V be a, not necessarily closed, subspace of $\mathcal{X}$. V is said to be frequency invariant if every x_0 in V has a (ξ, ω)–representation with $\xi(.) \in V(s)$.

Remark:

With the above result we have that if V is frequency invariant, then every x_0 in V has a (ξ, ω)–representation with $\xi(.) \in V_{-1}(s)$.

Example II.16.

Let (A, B) satisfy the assumption made on equation (2.1) and let λ be an arbitrary element of resolvent set of A and $u \in \mathcal{U}$. Define $V_{\lambda, u} := \text{span}\{(\lambda - A)^{-1} Bu\}$, then $V_{\lambda, u}$ is frequency invariant. By a simple calculation we obtain that $(s - A)(\lambda - A)^{-1} Bu = (s - \lambda)(\lambda - A)^{-1} Bu + Bu$, so

(2.15) $\qquad (\lambda - A)^{-1}Bu = (s - A)\left\{\dfrac{1}{(s-\lambda)}(\lambda - A)^{-1}Bu\right\} - B\left\{\dfrac{1}{(s-\lambda)}u\right\}$

Thus $V_{\lambda,u}$ is frequency invariant.

These subspaces together with the eigenvectors of A are the one–dimensional frequency invariant subspaces

$\qquad\qquad\qquad\qquad\qquad\qquad\qquad\qquad\qquad\qquad\qquad\qquad\qquad\quad$ □

Example II.17.

$\qquad$ Let (A,B) satisfy the assumption made on equation (2.1) and let C be a bounded operator from X to $\mathbb{R}^p$. Let V be the following subset of X.

(2.16) $\qquad V = \{x \in X | there\ exists\ an\ U(s) \in \mathcal{U}_{-1}(s)\ such\ that$

$$C(s-A)^{-1}BU(s) = C(s-A)^{-1}x\ for\ all\ s \geq s_x; s \in \rho(A)\}$$

We shall prove that this subspace is frequency invariant.

Define for $x \in V$ the function $\xi_x(s)$ to be $(s-A)^{-1}x - (s-A)^{-1}BU(s)$. Then $\xi_x(.)$ is continuous and for fixed s we have

(2.17) $\qquad C(\lambda - A)^{-1}\xi_x(s) = C(\lambda - A)^{-1}\left\{(s-A)^{-1}x - (s-A)^{-1}BU(s)\right\} =$

$$C(\lambda - A)^{-1}B\left\{\dfrac{-1}{(\lambda - s)}U(\lambda) + \dfrac{1}{(\lambda - s)}U(s)\right\}$$

So $\xi_x(.)$ is in V and by the definition of V we have that

(2.18) $\qquad\qquad\qquad x = (s-A)\xi_x(s) + BU(s);\quad s \geq s_x$

So V is frequency invariant.

$\qquad\qquad\qquad\qquad\qquad\qquad\qquad\qquad\qquad\qquad\qquad\qquad\qquad\quad$ □

In the sequel of this section we shall prove some technical properties of frequency invariant subspaces, which we shall need to prove the equivalence in section II.4. We advice the reader who is not interested in details to proceed to section II.4.

Lemma II.18 and lemma II.19 will imply that the (ξ,ω)–representation is unique if $Im\ B \cap V = \{0\}$. First we recall some notation of the previous section. $B^0(V)$ is a subspace of $Im\ B$ that has zero intersection with V and is of maximal dimension under this restriction, for short we shall write B^0.

Lemma II.18.

If a subspace V of X is frequency invariant, then every element of V has a (ξ, ω)–representation with $\xi(.) \in V_{-1}(s)$ and $B\omega(.) \in B_{-1}^0(s)$.

Proof:

By our assumption we know that every x in V has a (ξ, ω) representation such that $\lim\limits_{s \to \infty} s\omega(s)$ exists.

Let $b_1, .., b_{m_0}$ be a basis for $B^0(V)$, and let $b_{m_0+1}..., b_m$ be a basis for $B^1(V): = Im\ B \cap V$. Then, since V is a subspace , $b_1, .., b_m$ is a basis for $Im\ B$. Since V is frequency invariant and $B^1(V) \subset V$, we have that there exist pairs $(\ \xi_i(s), \omega_i(s)\ )$ such that $\xi_i(.) \in V_{-1}(s)$, $\omega_i(.) \in U_{-1}(s)$ and

$$(2.19) \qquad b_i = (s-A)\xi_i(s) - B\omega_i(s), \qquad i = m_0+1, .., m$$

If we write $\omega_i(s) = \begin{pmatrix} \omega_{i1}(s) \\ \vdots \\ \omega_{im}(s) \end{pmatrix}$, then (2.19) becomes:

$$(2.20) \qquad b_i = (s-A)\xi_i(s) - \sum_{j=1}^{m} b_j \omega_{ij}(s); \qquad i = m_0+1, .., m$$

We can write (2.18) in matrix notation, and obtain

$$(2.21) \qquad \begin{pmatrix} b_{m_0+1} \\ \vdots \\ b_m \end{pmatrix} = Q(s) \begin{pmatrix} b_{m_0+1} \\ \vdots \\ b_m \end{pmatrix} + R(s); \ where\ Q_{ij}(s) = -\omega_{ij}(s)$$

$$and\ R_i(s) = (s-A)\xi_i(s) - \sum_{j=1}^{m_0} b_j \omega_{ij}(s).$$

Since $\omega_i(.)$ is in $U_{-1}(s)$; $Q(s)$ is a continuous function and $\lim\limits_{s \to \infty} Q(s) = 0$. So $(I - Q(s))^{-1}$ exists for s sufficiently large, and rearranging equation (2.21) gives

(2.22)
$$\begin{pmatrix} b_{m_0+1} \\ \vdots \\ b_m \end{pmatrix} = (I - Q(s))^{-1} R(s) =$$

$$(I-Q(s))^{-1}(s-A)\begin{pmatrix} \xi_{m_0+1}(s) \\ \vdots \\ \xi_m(s) \end{pmatrix} - (I-Q(s))^{-1}\begin{pmatrix} \sum\limits_{j=1}^{m_0} b_j \omega_{m_0+1,j}(s) \\ \vdots \\ \sum\limits_{j=1}^{m_0} b_j \omega_{m,j}(s) \end{pmatrix} =$$

$$(s-A)(I-Q(s))^{-1}\begin{pmatrix} \xi_{m_0+1}(s) \\ \vdots \\ \xi_m(s) \end{pmatrix} - \sum\limits_{j=1}^{m_0} b_j (I-Q(s))^{-1}\begin{pmatrix} \omega_{m_0+1,j}(s) \\ \vdots \\ \omega_{m,j}(s) \end{pmatrix},$$

by simple linear algebra.

This last formula implies that each b_i, $m_0+1 \le i \le m$, has a (ξ, ω) representation with $\hat{\xi}_i(.)$ in $V_{-1}(s)$ and $B\hat{\omega}_i(.)$ in $B^0_{-1}(V)$, by taking $\hat{\xi}_i(s)$ to be the i-th row of $(I-Q(s))^{-1}\begin{pmatrix} \xi_{m_0+1}(s) \\ \vdots \\ \xi_m(s) \end{pmatrix}$ and $B\hat{\omega}_i(.)$ to be the i-th row of $\sum\limits_{j=1}^{m_0} b_j(I-Q(s))^{-1}\begin{pmatrix} \omega_{m_0+1,j}(s) \\ \vdots \\ \omega_{m,j}(s) \end{pmatrix}$. Furthermore since $Q(\infty)=0$, $\lim\limits_{s\to\infty} s\hat{\omega}_i(s) = \lim\limits_{s\to\infty} s\begin{pmatrix} \omega_{i,1}(s) \\ \vdots \\ \omega_{i,m_0}(s) \end{pmatrix}$, and by assumption this limit exists.

So we have proved the assertion of lemma II.18 for all elements in $Im\, B \cap V = B^1(V)$. Let x be an arbitrary element of V, then there exists a pair $(\xi(s), \omega(s))$, $\xi(.) \in V_{-1}(s)$ and $\omega(.) \in U_{-1}(s)$ such that

(2.23)
$$x = (s-A)\xi(s) - B\omega(s) = (s-A)\xi(s) - \sum\limits_{i=1}^{m_0} b_i \omega_i(s) - \sum\limits_{i=m_0+1}^{m} b_i \omega_i(s)$$

$$= (s-A)\xi(s) - \sum\limits_{i=m_0+1}^{m} \left[(s-A)\hat{\xi}_i(s) - B\hat{\omega}_i(s)\right]\omega_i(s) - \sum\limits_{i=1}^{m_0} b_i \omega_i(s) =$$

$$= (s-A)\left[\xi(s) - \sum\limits_{i=m_0+1}^{m} \hat{\xi}_i(s)\omega_i(s)\right] - \sum\limits_{i=m_0+1}^{m} \left[B\hat{\omega}_i(s)\right]\omega_i(s) - \sum\limits_{i=1}^{m_0} b_i \omega_i(s);$$

$\xi_i(s)$, $\hat{\xi}_i(s) \in V_{-1}(s)$ and $B\hat{\omega}_i(.) \in B^0_{-1}(s)$; $B^0 = span\{b_1, .., b_{m_0}\}$.

Thus x has a (ξ,ω) representation with $\tilde{\xi}(s) = \xi(s) - \sum\limits_{i=m_0+1}^{m} \hat{\xi}_i(s)\omega_i(s) \in V_{-1}(s)$,

$$B\tilde{\omega}(s) = \sum_{i=m_0+1}^{m}\left[B\hat{\omega}_i(s)\right]\omega_i(s) - \sum_{i=1}^{m_0} b_i\omega_i(s) \in B^0_{-1}(s) \text{ and } \lim_{s\to\infty} s\,\tilde{\omega}(s) \text{ exists.}$$

$\square$

lemma II.19.

If V is closed subspace that is frequency invariant, then there exists an $\hat{s} \in R$ such that $\left[(s-A)^{-1}B^0\right] \cap V = \{0\}$ for all s larger than $\hat{s}$.

Proof:

Let $b_1,...,b_{m_0}$ be a basis for B^0. Then by the Hahn–Banach theorem and the fact that $B^0 \cap V = \{0\}$, there exist functionals $\{g_i \in X', i=1,..,m_0\}$ such that $<g_i,b_j> = \delta_{ij}$ and $g_i|_V = 0$. Define for $s\in\rho(A)\cap R$ the following matrix valued function $S(s) := (<g_i, s(s-A)^{-1}b_j>)$. By the continuity of the resolvent operator we have that this function is continuous on $\rho(A)\cap R$. Furthermore we have that $S_{ij}(\infty) := \lim\limits_{s\to\infty} S_{ij}(s) = <g_i,b_j> = \delta_{ij}$. So $S(\infty)$ is nonsingular. Then by the continuity of $S(s)$ in plus infinity there exists an interval $[\hat{s},\infty)$ such that $S(s)$ is invertible on that interval.

On this interval we have that $(s-A)^{-1}B^0 \cap V = \{0\}$, since otherwise there would exists a vector $u^1 \neq 0$ and a $s_0 \geq \hat{s}$ such that $\sum\limits_{j=1}^{m_0} (s_0-A)^{-1}b_j u^1_j$ is contained in V, and so

$$S(s_0)u^1 = \begin{bmatrix} \sum\limits_{j=1}^{m_0} <g_1, s_0(s_0-A)^{-1}b_j u^1_j> \\ \cdot \\ \cdot \\ \cdot \\ \sum\limits_{j=1}^{m_0} <g_{m_0}, s_0(s_0-A)^{-1}b_j u^1_j> \end{bmatrix} = \begin{bmatrix} <g_1, s_0 \sum\limits_{j=1}^{m_0} (s_0-A)^{-1}b_j u^1_j> \\ \cdot \\ \cdot \\ \cdot \\ <g_{m_0}, s_0 \sum\limits_{j=1}^{m_0} (s_0-A)^{-1}b_j u^1_j> \end{bmatrix} =$$

$$= s_0 \begin{bmatrix} <g_1, \sum\limits_{j=1}^{m_0} (s_0-A)^{-1}b_j u^1_j> \\ \cdot \\ \cdot \\ <g_{m_0}, \sum\limits_{j=1}^{m_0} (s_0-A)^{-1}b_j u^1_j> \end{bmatrix} = \begin{bmatrix} 0 \\ \cdot \\ \cdot \\ \cdot \\ 0 \end{bmatrix}.$$

Thus $\det(S(s_0)) = 0$, providing the contradiction.

$\square$

The following lemma is the frequency domain version of lemma II.10.

Lemma II.20.

If a closed subspace V of X is frequency invariant, then every x in V has a *unique* (ξ, ω)-representation with $B\omega(.)$ in $B^0_{-1}(s)$ and $\xi(.)$ in $V_{-1}(s)$ and there exists an interval $[\hat{s}, \infty)$, which is independent of x, such that $\xi(s)$ and $\omega(s)$ are continuous on this interval, this $\hat{s}$ is the same as in lemma II.19.

Proof:

Let $\hat{s}$ be the constant of lemma II.19 and suppose that $g_i \in X'$ satisfies $g_i|_V = 0$ and $<g_i, b_j> = \delta_{ij}$, where $\{b_j\}_{j=1}^{m_0}$ is a basis for B^0. Furthermore let x be an arbitrary element of V, then by lemma II.18 it has a (ξ, ω) representation with $B\omega(.)$ contained in $B^0_{-1}(s)$. Thus on an interval $[\tau_x, \infty)$ x can be written as

$$(2.24) \qquad x = (s-A)\xi(s) - \sum_{i=1}^{m_0} b_i \omega_i(s), \quad \xi(.) \in V_{-1}(s)$$

(2.24) implies:

$$(2.25) \qquad (s-A)^{-1}x = \xi(s) - \sum_{i=1}^{m_0} (s-A)^{-1}b_i \omega_i(s)$$

Calculating $<g_i, (s-A)^{-1}x>; \; i=1,..,m_0$ gives

$$(2.26) \qquad <g_i, (s-A)^{-1}x> = <g_i, \xi(s)> - \sum_{i=1}^{m_0} <g_i, (s-A)^{-1}b_j> \omega_j(s)$$

$$= - \sum_{i=1}^{m_0} <g_i, (s-A)^{-1}b_j> \omega_j(s), \text{ since } \xi(s) \text{ is in } V.$$

If we premultiply (2.26) by s we obtain;

$$(2.27) \qquad <g_i, s(s-A)^{-1}x> = \sum_{i=1}^{m_0} <g_i, s(s-A)^{-1}b_j> \omega_j(s).$$

Or in matrix notation:

$$(2.28) \qquad \begin{bmatrix} <g_i, s(s-A)^{-1}x> \\ \\ <g_{m_0}, s(s-A)^{-1}x> \end{bmatrix} = -S(s) \begin{bmatrix} \omega_1(s) \\ \\ \omega_{m_0}(s) \end{bmatrix},$$

where $S_{ij}(s) = <g_i, s(s-A)^{-1}b_j>$.

Note that $S(s)$ and $<g_i, s(s-A)^{-1}x>$ are continuous on the interval

$[\hat{s},\infty)\subset\rho(A)$, and from lemma II.19 we have that $S(s)$ is invertible on this interval. So on $[\hat{s},\infty)$, $\{\omega_i(.)\}$, $i=1,..,m_0$ is the unique solution of equation (2.28). By (2.25) we have that there is only one choice for $\xi(s)$, that is:

$$(s-A)^{-1}x+(s-A)^{-1}\left(\sum_{i=1}^{m_0} b_i\omega_i(s)\right)$$

□

Let us remark that (2.28) implies that if $\mathcal{X}$ is finite dimensional, and so A is a matrix, then $\omega_i(s)$ is a rational function, and with the last line of the proof of lemma II.20 $\xi(s)$ is a rational function too. So if $\mathcal{X}$ is finite dimensional, then definition II.15 is the same as if we were to restrict ourselves to strictly proper rational functions, as in Hautus [19].

Before we can prove the equivalence between frequency and and closed loop invariance we need some properties of the set of all possible values of $\xi(s)$ for x in V, which we shall denote by Ξ_s.

Definition II.21: Ξ_s

If V is a frequency invariant subspace, then Ξ_{s_1} consists of all $\xi_1\in V\cap D(A)$ such that there exists a x in V with a (ξ,ω) representation; $x=(s-A)\xi(s)-B\omega(s)$, $\xi(.)\in V_{-1}(s)$, $\omega(.)\in U_{-1}(s)$, such that $\xi(s_1)=\xi_1$.

Lemma II.22.

If V is a closed subspace that is frequency invariant and $Im\,B\cap V=\{0\}$, then we have that there exists a real $\hat{s}$ such that for any $s_1\geq\hat{s}$, the equalities

(2.29)
$$\begin{cases} x=(s_1-A)\xi_1-B\omega_1 \quad and \\ \\ x=(s_1-A)\xi_2-B\omega_2, \qquad\qquad when\ x,\ \xi_1\ and\ \xi_2\ in\ V, \end{cases}$$

imply that $\xi_1=\xi_2$.

Proof:

Since V is a closed subspace of $\mathcal{X}$ and $Im\,B\cap V=\{0\}$, this lemma is a simple corollary of lemma II.19, and $\hat{s}$ in this lemma is the same as in lemma II.19

□

Lemma II.23.

Let V be a closed frequency invariant subspace with $Im\, B \cap V = \{0\}$, then:

$$\Xi_{s_1} = \Xi_{s_2},$$

for all $s_1, s_2 \in [\hat{s}, \infty)$, where $\hat{s}$ is as in lemma II.19.

Proof:

Let ξ_1 be an element of Ξ_{s_1}, then there exists a x in V with

$$x = (s_1 - A)\xi_1 - B\omega(s_1); \quad \xi_1 = \xi(s_1)$$

Rewriting this equation gives $x = (s_1 - s_2 + s_2 - A)\xi_1 - B\omega(s_1)$, or

(2.30) $$(s_2 - s_1)\xi_1 + x = (s_2 - A)\xi_1 - B\omega(s_1)$$

$(s_2 - s_1)\xi_1 + x$ is an element of V thus it has a (ξ, ω) representation. So there exists a pair $(\hat{\xi}, \hat{\omega})$ such that

(2.31) $$(s_2 - s_1)\xi_1 + x = (s - A)\hat{\xi}(s) - B\hat{\omega}(s)$$

From lemma II.20 we have that equation (2.31) holds on $[\hat{s}, \infty)$. Now relations (2.30) and (2.31) with lemma II.22 imply that $\hat{\xi}(s_2) = \xi_1$. So $\Xi_{s_1} \subset \Xi_{s_2}$. By symmetry we conclude that $\Xi_{s_1} = \Xi_{s_2}$.

$\square$

Lemma II.24.

Let V be a closed frequency invariant subspace with $Im\, B \cap V = \{0\}$, then $\Xi := \Xi_{\hat{s}}$ is closed in the graph norm of A, for definition see Davies [14, lemma 1.6.], where $\hat{s}$ is defined as in lemma II.19.

Proof:

Let ξ_n be a sequence in Ξ such that $\xi_n \to y$ and $A\xi_n \to z$. Since A is a closed operator, we have that $y \in D(A)$ and $Ay = z$. Let $\{b_1, .., b_{m_0}\}$ be a basis for $Im\, B$, then since $Im\, B \cap V = \{0\}$ and V is a closed subspace there exist $g_i \in X'$ such that $g_i\big|_V = 0$ and $<g_i, b_j> = \delta_{ij}$. Since ξ_n is an element of Ξ, there exist x_n in V and ω_n in $\mathcal{U}$ such that

(2.32) $$x_n = (\hat{s} - A)\xi_n - B\omega_n = (\hat{s} - A)\xi_m - \sum_{j=1}^{m_0} b_j\omega_{nj}$$

Since $x_n \in V$, we have that

$0 = \;<g_i, x_n> \;= \;<g_i, (\hat{s} - A)\xi_n - B\omega_n> \;= - <g_i, A\xi_n> - \;<g_i, b_i> \omega_{ni}$

and thus $\quad <g_i, A\xi_n> \;= -\omega_{ni}$

So ω_{ni} converges as $n \to \infty$, $i = 1,..,m_0$. Thus ω_n converges to say $\omega \in \mathcal{U}$ and since $x_n = (\hat{s} - A)\xi_n - B\omega_n$ we have that x_n converges to x. Since V is closed we have that $x \in V$ and so there exist $\xi(s)$ and $\omega(s)$ such that

$$x = (s - A)\xi(s) - B\omega(s).$$

By definition x is also equal to $(\hat{s} - A)y - B\omega$. From lemma II.22 we have that $y = \xi(\hat{s})$, and thus $y \in \Xi$.

$\qquad\qquad\qquad\qquad\qquad\qquad\qquad\qquad\qquad\qquad\qquad\qquad\qquad$ □

Section II.4: Equivalence

In the theory of system invariant subspaces the equivalence between closed loop and open loop invariance is of great importance. This equivalence tells us that, if we can find an input such that the trajectory stays in a subspace, then we can also find a feedback law such that the trajectory stays in this subspace. In Basile & Marro [1] and Wonham [42] this equivalence was proved in the case that the state space is finite dimensional. If the state space is infinite dimensional, then the equivalence was proved by Schmidt and Stern [32] provided that A is bounded. However, the interesting case in infinite dimensions is when A is unbounded, but generates a C_0–semigroup; and this is the focus of this section. We formulate and prove this equivalence for the system (2.1). Furthermore we shall prove that closed loop invariance is equal to frequency invariance for closed linear subspaces and that the equivalence between open and closed loop invariance is lost if the subspace is not closed.

We shall begin by showing that there is equivalence between (A,B)– and feedback invariance. However there is in general no equivalence between (A,B)– and closed loop invariance even if we impose the extra condition that $V \cap D(A)$ is dense in V, as shown in Schmidt and Stern [32].

Lemma II.25.

If $V_1 \subset D(A)$ is a linear subspace, closed with respect to the graph norm of A and $V_2 \subset \mathcal{X}$ is a closed linear subspace with

(2.33) $$AV_1 \subset V_2 + Im\, B,$$

then there exists an A-bounded feedback law F such that

(2.34) $$(A + BF)V_1 \subset V_2$$

Proof:

If $\mathcal{X}$ is finite dimensional, then the proof can be found in Basile & Marro [1] and Wonham [42, p. 88]. The general proof that will be presented here is an adaptation of the proof given by Pandolfi [27].

Define $\mathcal{X}_A$ to be the graph of A, with the graph norm $\|(x, Ax)\|_A = \|x\| + \|Ax\|$, where $\|.\|$ is the norm of $\mathcal{X}$. A is a bounded operator from $\mathcal{X}_A$ to $\mathcal{X}$, and V_1 is by definition closed in $\mathcal{X}_A$.

If $v_1 \in V_1$, then there exists $v_2 \in V_2$ and $u \in \mathcal{U}$ such that $Av_1 = v_2 + Bu$, and u and v_2 are uniquely determined if we assume that $u \in [Ker\, B]^\perp$ (the annihilator of $Ker\, B$) and $Bu \in B^0(V_2)$.

Let F be defined by $Fv_1 = -u; \ \forall\ v_1 \in V_1$. The operator F is linear, since u is uniquely determined. We shall show that F is a closed operator in $\mathcal{X}_A$. Let us assume that (v_1^n, Fv_1^n) converges to $(v_1, -u)$, thus $v_1^n \to v_1$ and $Av_1^n \to Av_1$. We must prove that $Fv_1 = -u$. This is obvious since

$$w^n := Av_1^n - Bu^n = Av_1^n + BFv^n \text{ converges to } Av_1 - Bu =: w,$$

and $v_1 \in V_1$ since V_1 is closed and $Bu \in B^0(V_2)$, since $B^0(V_2)$ is closed.

Thus F is a closed operator from the whole of V_1, with induced norm of $\|.\|_A$, to $\mathcal{U}$. By the Closed Graph Theorem F is a bounded operator on V_1, with norm $\|.\|_A$, by the Hahn-Banach Theorem and the fact that, since $\mathcal{U}$ is finite dimensional, F has finite dimensional range, F has a bounded extension on $\mathcal{X}_A$. From Kato [22, p. 191 and 245] we have that F is A-bounded on $\mathcal{X}$.   □

Theorem II.26.

If V is a closed subspace of $\mathcal{X}$, then (A, B) and feedback invariance are equivalent.

Proof:

This is a easy corollary of lemma II.25 since if V is a closed subspace and A is a closed operator, then $V \cap D(A)$ is closed with respect to the graph norm of A.

$\square$

Now we have proved all the ingredients for our main result.

Theorem II.27.

Let V be a <u>closed</u> linear subspace of X, then the following concepts of invariance are equivalent:

a) V is closed loop invariant

b) V is open loop invariant

c) V is frequency invariant

We remark that lemma I.4 can be seen as a special case of this theorem i.e. no control action thus $B = 0$.

If X is finite dimensional, then the equivalence between these concepts are known, see for $a) \leftrightarrow b)$ e.g. Basile & Marro [1] and for $a) \leftrightarrow c)$ Hautus [19]. Since we have equivalence between these invariance concepts we introduce a new concept that we shall use if a subspace satisfies II.27 $a),b)$, or $c)$.

Definition II.28 Controlled Invariance

A closed linear subspace V of X is called controlled invariant if it satisfies II.27 $a),b)$ or, equivalently $c)$.

As we shall see in the next chapter the equivalence between open loop and closed loop invariance as well as the equivalence between frequency and closed loop invariance is lost in general if the subspace V is not closed. From the previous chapter we see that there is no hope for equivalence between controlled and (A,B) invariance in general. However there is one case were this equivalence holds.

Lemma II.29.

If V is a closed subspace contained in the domain of A, then the following assertions are equivalent:

a) V is closed loop invariant

b) V is open loop invariant

c) V is frequency invariant

d) V is (A,B) invariant

e) V is feedback invariant

Proof:

The equivalence between a) and e) is a consequence of lemma I.7, the other equivalences follow from theorem II.26 and II.27.

We shall prove theorem II.27 by showing a) $\Rightarrow$ b) $\Rightarrow$ c) $\Rightarrow$ a).

Proof of Theorem II.27:

a) => b)

Let F be the feedback law such that $T_{A+BF}(t)V \subset V$.

Defining $x(t)$ as $T_{A+BF}(t)x_0$ and $u(t)$ as $FT_{A+BF}(t)x_0$, then Curtain and Pritchard [9, th. 2.31] gives the desired result.

b) => c)

Let x_0 be an element of V. Then from lemma II.11 we have that there exists an $u(.) \in C([0,\infty);\mathcal{U})$ such that

$$(2.35) \qquad x(t) = T_A(t)x_0 + \int_0^t T_A(t-s)Bu(s)ds$$

is in V and $\|u(t)\| \leq Me^{\alpha t}$, for some M and α in $\mathbf{R}$. The exponential boundedness of $u(.)$ implies the same for $x(.)$, and so we can take the Laplace transform of equation (2.35). Define $\xi(.)$ to be the Laplace transform of $x(t)$ and $\omega(.)$ the Laplace transform of $u(t)$. Then equation (2.35) gives on an interval $[r_{x_0},\infty)$ the following relation

$$(2.36) \qquad \xi(s) = (s-A)^{-1}x_0 + (s-A)^{-1}B\omega(s)$$

Since V is a closed subspace we have that $\xi(s) \in V$, and with theorem 8.6-1 of Zemanian [46] we have that $\lim_{s \to \infty} s\omega(s) = \lim_{t \downarrow 0} u(t) = u(0)$. So $\xi(.)$, $\omega(.)$ is a (ξ,ω) representation of x_0.

c) => a)

Since V is a closed subspace, we see, by lemma II.20, that we may restrict our input operator B to $\tilde{B}$, such that $\tilde{B}$ is injective, $B^0(V) = \tilde{B}^0(V)$ and $\tilde{B}^1(V) = Im\ \tilde{B} \cap V = \{0\}$ (see (2.8)). Then V is also frequency invariant for the system $(A,\tilde{B})$. So we may assume without loss of generality that $Im\ B \cap V = \{0\}$.

By assumption we have that all x in V admit the following decomposition.

(2.37) $x = (s - A)\xi(s) - B\omega(s).$

If s is larger than $\hat{s}$ (see lemma II.19 and definition II.21), then since $\xi(s) \in \Xi_s$, (2.37) implies that

(2.38) $A\Xi_s \subset V + Im\ B.$

From lemma II.24 we have that $\Xi_s = \Xi$. So (2.38) implies that

(2.39) $A\Xi \subset V + Im\ B.$

Ξ is closed in the graph norm of A, so from lemma II.25 we have the existence of an A-bounded F such that

$$(A + BF)\Xi \subset V.$$

Rearranging equation (2.37) gives

$$x = (s - A)\xi(s) - B\omega(s) = (s - A - BF)\xi(s) - B(\omega(s) - F\xi(s))$$

Thus $B(\omega(s) - F\xi(s)) \in V$, so by the assumption made in the beginning of this proof $\omega(s) = F\xi(s)$. So

(2.40) $x = (s - A - BF)\xi(s)$

It remains to prove that this feedback law is bounded. We shall begin by showing that the mapping $x \mapsto s\omega(s)$ is a bounded operator from V to $\mathcal{U}$ for $s \in [\hat{s}, \infty) \cap \rho(A)$. Let for $s \in [\hat{s}, \infty) \cap \rho(A)$, F_s be defined as $F_s x := s\omega(s)$, where $\omega(s)$ is the (unique) input from equation (2.37). It is easy to show that these operators are closed, and thus, since they are defined on the closed subspace V, bounded.

By the definition of $\mathcal{U}_{-1}(s)$ we have that $\lim_{s\to\infty} s\omega(s)$ exists, so for every x in V we can define the operator $\tilde{F}x$ by $\tilde{F}x = \lim_{s\to\infty} F_s x = \lim_{s\to\infty} s\omega(s)$. Then by the uniqueness of $\omega(.)$ $\tilde{F}$ is a linear operator defined on V. By the Uniform Boundedness theorem, and since V is closed we have that $\tilde{F}$ is a bounded operator on V. The Hahn Banach Theorem gives that $\tilde{F}$ can be extended as a bounded operator to the whole of $\mathcal{X}$. We shall show that on Ξ $F = \tilde{F}$. Let ξ_0 be an element Ξ, then there exists a x_0, $\xi(.)$ and $\omega(.)$ such that

$$(2.41) \qquad x_0 = (s-A)\xi(s) - B\omega(s) \text{ and } \xi(\hat{s}) = \xi_0$$

Rearranging this equation gives that

$$(2.42) \qquad \xi_0 = (s-A)\left[\frac{\xi(s)-\xi_0}{\hat{s}-s}\right] - B\left[\frac{\omega(s)-\omega(\hat{s})}{\hat{s}-s}\right]$$

Define $\xi_1(.)$ and $\omega_1(.)$ to be respectively $\left[\dfrac{\xi(s)-\xi_0}{\hat{s}-s}\right]$ and $\left[\dfrac{\omega(s)-\omega(\hat{s})}{\hat{s}-s}\right]$.

Then $\tilde{F}\xi_0 = \lim_{s\to\infty} s\omega_1(s) = \omega(\hat{s}) = F\xi(\hat{s}) = F\xi_0$. So on Ξ F is equal to the bounded operator $\tilde{F}$. So equation (2.40) implies that for sufficiently large s $(s-A-BF)^{-1}x = \xi(s)\in V$, and lemma I.4 concludes the proof. $\qquad\square$

CHAPTER III: DISTURBANCE DECOUPLING PROBLEM

In this chapter we shall consider the disturbance decoupling problem (DDP): given the system

(3.1) $\qquad \dot{x}(t) = Ax(t) + Bu(t) + Eq(t), \quad z(t) = Dx(t),$

where A and B are the same as in (2.1) and E and D are bounded linear operators from respectively Q to X and X to Z, find an bounded feedback law such that, in the closed loop system, $z(.)$ does not depend on $q(.)$.

So pictorially we have the following situation;

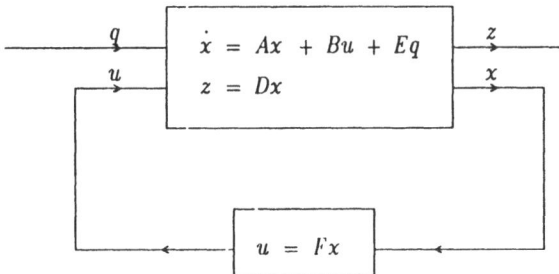

Thus the Disturbance Decoupling Problem is to design a feedback law F such that the transfer from q to z is zero i.e. $D(s - A - BF)^{-1}E \equiv 0$.

The next theorem gives the link between controlled invariance and DDP.

Lemma III.1.

The Disturbance Decoupling Problem is solvable if and only if there exists a controlled invariant subspace V such that $Im\ E \subset V \subset Ker\ D$.

Proof:

See Curtain [6]. □

In the finite dimensional theory it turns out that the largest controlled invariant subspace contained in the kernel of D can always be calculated. Here we shall also introduce this subspace, but the calculation is a difficult problem in general.

Definition III.2: $V^*(K)$.

Let K be a closed subspace in X. Then we shall denote by $V^*(K)$ the largest controlled invariant subspace contained in K.

If this subspace exists, then we have the following nice result.

Theorem III.3.

If $V^*(Ker\ D)$ exists, then DDP is solvable if and only if

(3.2) $$V^*(Ker\ D) \supset Im\ E.$$

Proof:

See Curtain [6]

□

In Curtain [6] the question of sufficient conditions for the existence of $V^*(K)$ is posed. We remark here that $V^*(K)$ need not always exist, see Pandolfi [27] for a counter example for delay equations, or see appendix E for a counter example for partial differential equations. Note that if $V^*(K)$ exists, then it must be closed, since $T_{A+BF}(t)V \subset V$ implies that $T_{A+BF}(t)\bar{V} \subset \bar{V}$.

The Disturbance Decoupling problem will be investigated in the frequency–domain as well as in the time–domain in section III.1 and III.2 respectively. In section III.3 we investigate some properties of closed loop invariant subspaces using the results from III.1 and III.2.

Section III.1: DDP in Frequency Domain

Keeping the equivalence between closed loop– and frequency invariance in mind we define the natural candidate for $V^*(K)$.

Definition III.4: $V_{\Sigma}(K)$

We assume that K is a closed linear subspace.

Let $V_{\Sigma}(K)$ be the subset of X which contains all $x \in X$ with a (ξ,ω) representation with $\xi(.)$ in $K(s)$, (see definitions II.12 and II.13).

The next lemma will show that $V_{\Sigma}(K)$ is the supremal frequency invariant subspace contained in the closed subspace K.

Lemma III.5

a) Every frequency invariant subspace contained in K is contained in $V_{\Sigma}(K)$.

b) Every closed loop invariant subspace contained in K is contained in $V_{\Sigma}(K)$.

c) $V_{\Sigma}(K) \subset K$.

d) $V_{\Sigma}(K)$ is the supremal frequency invariant subspace, contained in K.

Proof:

a) Obvious, with the definition of $V_{\Sigma}(K)$ and the definition of frequency invariance.

b) Obvious, with theorem II.27, since if V is closed loop invariant, then $\overline{V}$ is it too.

c) Let $(\xi(s), \omega(s))$ be a (ξ, ω) representation of x in $V_{\Sigma}(K)$, then $\xi(s) \in K$ and with lemma II.14 and the fact that K is closed we conclude that $x = \lim_{s \to \infty} s\xi(s)$ is an element of K.

d) Let x be an element of $V_{\Sigma}(K)$, so that there exist strictly proper functions $\xi(s)$ and $\omega(s)$ which are continuous on an interval $[\tau_x, \infty)$ and such that $\xi(s)$ is in K and

(3.3) $\qquad x = (s - A)\xi(s) - B\omega(s); \quad \text{for } s > \tau_x$

(see definition II.15). Let s_0 be an arbitrary, but fixed point in $\mathbb{R}$ with $s_0 > \tau_x$, then by using (3.3) we get

(3.4) $\qquad 0 = (s - A)\xi(s) + (s_0 - A)(-\xi(s_0)) - B\omega(s) + B\omega(s_0) \quad =$

$$(s - A)\Big[\xi(s) - \xi(s_0)\Big] + (s - s_0)\xi(s_0) - B\Big[\omega(s) - \omega(s_0)\Big]$$

So if $s \neq s_0$, then (3.4) implies that

(3.5) $\qquad \xi(s_0) \; = \; (s - A)\left[\dfrac{(\xi(s) - \xi(s_0))}{s_0 - s}\right] - B\left[\dfrac{(\omega(s) - \omega(s_0))}{s_0 - s}\right]$

Define for $s > s_0$ $\quad \xi_1(s) := \left[\dfrac{(\xi(s) - \xi(s_0))}{s_0 - s}\right]$ and $\omega_1(s) := \left[\dfrac{(\omega(s) - \omega(s_0))}{s_0 - s}\right]$.

It is not hard to show that (ξ_1, ω_1) is a (ξ, ω)–representation of

$\hat{x} = \xi(s_0)$ with $\lim\limits_{s \to \infty} sw_1(s) = \omega(s_0)$. Since $\xi(s)$ is in K, $\xi_1(s)$ is in K. Thus by the definition of $V_\Sigma(K)$, $\xi(s_0)$ is in $V_\Sigma(K)$. Since s_0 was an arbitrary element of $[r_x, \infty)$ we may conclude that x has a (ξ, ω)–representation with $\xi(.) \in V_\Sigma(K)$. Now since x was an arbitrary element of $V_\Sigma(K)$ we have that $V_\Sigma(K)$ is frequency invariant.

By c) we obtain $V_\Sigma(K) \subset K$ and a) implies that $V_\Sigma(K)$ is the largest frequency invariant subspace with this property.

□

With this lemma we can easily prove the next theorem.

Theorem III.6.

Let K be a closed subspace of X.

If $V_\Sigma(K)$ is closed, then $V^*(K)$ exists and is equal to $V_\Sigma(K)$.

Proof:

If $V_\Sigma(K)$ is closed, then by lemma III.5.d) and theorem II.27 we conclude that $V_\Sigma(K)$ is closed loop invariant. Furthermore from lemma III.5.a) and c) it must be the largest with this property. So $V^*(K)$ exists and is equal to $V_\Sigma(K)$.

□

Remark:

We remark that the converse of the theorem does not hold, i.e. if $V^*(K)$ exists, then $V_\Sigma(K)$ need not be closed, see example E.10.

So the question which arises is under what conditions is $V_\Sigma(K)$ closed. It turns out that these conditions can be formulated in terms of $Im\,B$ and $V_\Sigma(K)$, which we do in the next lemma. In this lemma we shall use the following notation.

If V is a subspace of X, then $V^\perp \subset X'$ will denote the annihilator of V i.e. $V^\perp = \{f \in X' |$ such that $f(V) = 0\}$. Furthermore if $\{f_n\}$ is a sequence in X', then this sequence is weak * convergent to f if $<f_n, x> \to <f, x>$ for all $x \in X$, where $<f, x>$ denotes the operation of the functional f on x. See for more details Kato [22, p. 136] and Yosida [44].

Lemma III.7.

Let $B^0(V_\Sigma(K))$ denote the largest subspace of $Im\,B$ that has zero intersection with $V_\Sigma(K)$, let $\{b_1,..,b_{m_0}\}$ be a basis of $B^0(V_\Sigma(K))$ and let $B^1(V)$ be $Im\,B\cap V$. Then the following assertions are equivalent.

i) $V_\Sigma(K)$ is closed.

ii) $\left(V_\Sigma(K)\right)^\perp \cap D(A')$ is weak * dense in $\left(V_\Sigma(K)\right)^\perp$ and $B^1(V_\Sigma(K)) = B^1(\overline{V_\Sigma(K)})$

iii)   There exist functionals $\{f_i\}_{i=1}^{m_0}$ in $\left(V_\Sigma(K)\right)^\perp \cap D(A')$ such that $<f_i, b_j> = \delta_{ij}$, where $D(A')$ is the domain of the dual operator of A, see Yosida [44].

Remark:

Before proving this theorem, we remark that the result of this lemma is independent of the basis of $B^0(V_\Sigma(K))$ and of the actual choice of $B^0(V_\Sigma(K))$. Furthermore it is clear by simple linear algebra that the condition in the third assertion of this lemma is equivalent to the existence of functionals $f_i \in \left(V_\Sigma(K)\right)^\perp \cap D(A')$ such that the matrix $S_\infty = <f_i, b_j>$ is nonsingular.

Proof of lemma III.7:

i)=>ii)

So we must show that the annihilator of $V_\Sigma(K)$, $V_\Sigma(K)^\perp$ intersected with $D(A')$ is sufficiently rich.

If $V_\Sigma(K)$ is a closed subspace, then by theorem III.6 it is closed loop invariant. So by lemma I.4.c) there exists a bounded feedback law such that $(\lambda - A - BF)^{-1} V_\Sigma(K) \subset V_\Sigma(K)$ for all $\lambda \in \mathbb{R}$ sufficiently large.

Let y be an arbitrary element of $\left(V_\Sigma(K)\right)^\perp$, then we shall show that $(\lambda - A' - F'B')^{-1} y$ is an element of $\left(V_\Sigma(K)\right)^\perp \cap D(A')$ and $\lambda(\lambda - A' - F'B')^{-1} y$ is weak * convergent to y, $\lambda \to \infty$. This will prove that is $\left(V_\Sigma(K)\right)^\perp \cap D(A')$ weak * dense in $\left(V_\Sigma(K)\right)^\perp$.

Let x be an arbitrary element of $V_\Sigma(K)$, then since $V_\Sigma(K)$ is $(\lambda - A - BF)^{-1}$ invariant we have that

$$< (\lambda - A' - F'B')^{-1} y, x > = <y, (\lambda - A - BF)^{-1} x > = 0$$

Thus $(\lambda - A' - F'B')^{-1}y$ is an element of $\big(V_{\Sigma}(K)\big)^{\perp} \cap D(A')$.

Now let x be an arbitrary element of $\mathcal{X}$, then

$$\lim_{\lambda \to \infty} <\lambda(\lambda - A' - F'B')^{-1}y, x> = \lim_{\lambda \to \infty} <y, \lambda(\lambda - A - BF)^{-1}x> = <y, x>.$$

This proves the assertion.

ii)=>iii)

Since $B^1(V_{\Sigma}(K)) = B^1(\overline{V_{\Sigma}(K)})$ we have from the well known Hahn–Banach theorem the existence of functionals $\{g_i\}_{i=1}^{m_0}$ in $\big(V_{\Sigma}(K)\big)^{\perp}$ such that the matrix $S' = <g_i, b_j>$ is non singular. Since $\big(V_{\Sigma}(K)\big)^{\perp} \cap D(A')$ is weak * dense in $\big(V_{\Sigma}(K)\big)^{\perp}$, there also exist functionals $\{f_i\}_{i=1}^{m_0} \subset \big(V_{\Sigma}(K)\big)^{\perp} \cap D(A')$ such that $S = <f_i, b_j>$ is non singular. Taking linear combinations of these f_i's gives the desired result.

iii)=>i)

$V_{\Sigma}(K)$ is frequency invariant and by lemma II.18, we obtain that every x in $V_{\Sigma}(K)$ has a (ξ, ω) representation with $B\omega(s) = \sum_{i=1}^{m_0} b_i \omega_i(s)$. Using the same argument as in the proof of lemma II.20 we can calculate this (ξ, ω) representation. So $\omega_i(s)$ is given by (2.27)

$$(3.6) \qquad \begin{pmatrix} \omega_1(s) \\ \vdots \\ \omega_{m_0}(s) \end{pmatrix} = -S(s)^{-1} \begin{pmatrix} <f_1, s(s-A)^{-1}x> \\ \vdots \\ <f_{m_0}, s(s-A)^{-1}x> \end{pmatrix} \qquad and$$

$$\xi(s) = (s-A)^{-1}x + (s-A)^{-1}\left[\sum_{i=1}^{m_0} b_i \omega_i(s) \right],$$

where $S_{ij}(s) = <f_i, s(s-A)^{-1}b_j>$

Let x be a element of $\overline{V_{\Sigma}(K)}$, then there exists a sequence $\{x_n\}$ in $V_{\Sigma}(K)$ which converges to x. x_n is an element of $V_{\Sigma}(K)$ so it has a (ξ, ω) representation with $\omega^n(s) = \begin{pmatrix} \omega_1^n(s) \\ \cdot \\ \omega_{m_0}^n(s) \\ \cdot \\ \omega_m^n(s) \end{pmatrix}$, where m is the dimension of $Im\, B$. With the above we may choose $w_{m_0+1}^n(s) = ,.., = w_m^n(s) = 0$ and $\begin{pmatrix} w_1^n(s) \\ \vdots \\ w_{m_0}^n(s) \end{pmatrix}$ as in (3.6) with

x replaced by x_n. With this choice it can easily be seen that if x_n converges to x, then $w^n(s)$ converges, and $w_i(s) := \lim_{n \to \infty} w_i^n(s) = 0$ if $i > m_0$ and $\begin{pmatrix} w_1(s) \\ \vdots \\ w_{m_0}(s) \end{pmatrix}$ is the right hand side of (3.6). From (3.6) it is easily seen

that $w(s)$ is continuous on an interval $[r, \infty)$ and since for all $x \in \overline{V_\Sigma(K)}$,

$<f_i, s(s-A)^{-1}x> = <f_i, x> + <f_i, A(s-A)^{-1}x> = 0 + <A'f_i, (s-A)^{-1}x>$ we have that

$$\lim_{s \to \infty} sw(s) = \lim_{s \to \infty} S(s)^{-1} \lim_{s \to \infty} \begin{pmatrix} <A'f_1, s(s-A)^{-1}x> \\ \vdots \\ <A'f_{m_0}, s(s-A)^{-1}x> \end{pmatrix} = \begin{pmatrix} <A'f_1, x> \\ \vdots \\ <A'f_{m_0}, x> \end{pmatrix},$$

since $\lim_{s \to \infty} S_{ij}(s) = \delta_{ij}$

Define $\xi(s) = (s-A)^{-1}x - (s-A)^{-1}Bw(s)$; then $\xi(s)$ is continuous, $x = (s-A)\xi(s) - Bw(s)$ and $\xi(s)$ is the limit of $\xi^n(s) = (s-A)^{-1}x_n - (s-A)^{-1}Bw^n(s)$, as $n \to \infty$, thus $\xi(s)$ is in K.

So x has a (ξ, w) representation, with $\xi(.)$ in $K(s)$, thus x is in $V_\Sigma(K)$. $\qquad\square$

The importance of this theorem lies in its usefulness in deriving further results, not in its direct application to specific systems. Even for the case of A bounded it is not clear if the conditions are fulfilled. However in the next lemma we show that if A is a bounded operator, then $V_\Sigma(K)$ is closed.

Lemma III.8.

If A is a bounded linear operator, then $V^*(K)$ exists and it is equal to $V_\Sigma(K)$.

Proof:

Suppose first that the following holds:

(3.7) $\qquad A\left[\overline{V_\Sigma(K)}\right] \subset \overline{V_\Sigma(K)} + Im\, B.$

Then with Schmidt and Stern [32], $\overline{V_\Sigma(K)}$ is controlled invariant. So with lemma III.5 $\overline{V_\Sigma(K)} = V^*(K)$, but $V^*(K) \subset V_\Sigma(K)$, so $V_\Sigma(K) = \overline{V_\Sigma(K)}$. Thus it remains to prove (3.7).

Let x be an element of $V_\Sigma(K)$, then there exists $\xi(.) \in K(s)$ and

$\omega(.)\in\mathcal{U}_{-1}(s)$ such that

(3.8) $$x = (s-A)\xi(s) - B\omega(s)$$

With lemma II.14 we have that $x = \lim_{s\to\infty} s\xi(s)$. Rearranging equation (3.8) gives

$$As\xi(s) = s^2\xi(s) - sx - Bs\omega(s)$$

From lemma III.5.d) it follows that $\xi(s)\in V_{\Sigma}(K)$.

So $As\xi(s)\in V_{\Sigma}(K)+\overline{Im\ B}\subset V_{\Sigma}(K)+\overline{Im\ B}$. Since A is a bounded operator and

$s\xi(s)\to x$ as $s\to\infty$, we have that $Ax\in V_{\Sigma}(K)+\overline{Im\ B}=V_{\Sigma}(K)+Im\ B$, because $Im\ B$ is finite dimensional. Thus

$$AV_{\Sigma}(K)\subset V_{\Sigma}(K)+Im\ B.$$

Using once again the fact that A is a bounded operator and $\overline{V_{\Sigma}(K)}+Im\ B$ is a closed subspace we have proved that $A\left[\ \overline{V_{\Sigma}(K)}\ \right]\subset\overline{V_{\Sigma}(K)}+Im\ B$ and thus this lemma.

$\square$

The next lemma will give sufficient conditions for theorem III.6, which are easy to verify. These conditions first occurred in Curtain [6], where the aim was to give sufficient conditions for the existence of $V^*(K)$.

Lemma III.9.

Let $Ker\ D$ be the kernel of a bounded linear operator D. Then $V_{\Sigma}(Ker\ D)$ is closed if either of the following conditions holds.

a) There exists a q in $\mathbb{N}\cup\{0\}$ such that DA^i has a bounded extension defined on the whole of $\mathcal{X}$ (denoted by $\overline{DA^i}$) for $0\le i\le q+1$, $\overline{DA^iB}=0$ for $0\le i<q$ and $\overline{DA^qBu}\ne 0$ for all $u\ne 0$ in $\mathcal{U}$.

b) $D(s-A)^{-1}B=0$ for all s in $[\hat{s},\infty)$ for some $\hat{s}$ in $\mathbb{R}$.

Proof:

a)

Suppose that $x\in V_{\Sigma}(Ker\ D)$ and condition a) holds. From lemma II.14 we have that $x = \lim_{s\to\infty} s\xi(s)$ if

(3.9) $x = (s-A)\xi(s) - B\omega(s).$

Using the fact that $\omega(s)$ is strictly proper, lemma II.14 and equation (3.9) we obtain

(3.10) $\lim_{s\to\infty} A\xi(s) = 0.$

If we premultiply (3.9) by D we get since $\xi(s) \in \text{Ker } D$ and $DB = 0$

(3.11) $Dx = D(s-A)\xi(s) - DB\omega(s) = sD\xi(s) - DA\xi(s) - DB\omega(s) = -DA\xi(s)$

Using the fact that $A\xi(s)$ is strictly proper and D is a bounded operator we get $Dx = 0$ and $DA\xi(s) = 0$. By induction it is easy to prove that $\overline{DA^i}x = 0$ and $\overline{DA^{i+1}}\xi(s) = 0$ for $0 \le i < q$.

Now we shall consider $i = q$

(3.12) $\overline{DA^q}x = \overline{DA^q}(s-A)\xi(s) - \overline{DA^q}B\omega(s) = s\overline{DA^q}\xi(s) - \overline{DA^q}(A\xi(s))$

$-\overline{DA^q}B\omega(s) = 0 - \overline{DA^q}(A\xi(s)) - \overline{DA^q}B\omega(s).$

Using the strict properness of $A\xi(s)$ and $\omega(s)$ we get that $\overline{DA^q}x = 0$ for all x in $V_\Sigma(\text{Ker } D)$. With the fact that $\overline{DA^q}$ is a bounded operator we obtain that $\overline{DA^q}x = 0$ for all $x \in \overline{V_\Sigma}(\text{Ker } D)$.

Let z be an element of the dual space of Z, with this functional we can define a bounded functional on X, i.e. $<z, \overline{DA^q}.>$. This functional is zero on $V_\Sigma(\text{Ker } D)$ and since DA^{q+1} has a bounded extension $\overline{DA^q}$. is an element of $D(A')$. The uniformity of the zero structure gives that one can find z_i such that $<z_i, \overline{DA^q}b_j> = \delta_{ij}$, where $\{b_j\}_{j=1}^m$ is a basis for $\text{Im } B$. Thus by lemma III.7 $V_\Sigma(\text{Ker } D)$ is closed.

b)

In the case that $D(s-A)^{-1}B = 0$, $V_\Sigma(\text{Ker } D)$ is equal to the subset in X of all x in $\text{Ker } D$ with $D(s-A)^{-1}x = 0$. This follows since for x in $V_\Sigma(\text{Ker } D)$,

$(s-A)^{-1}x = \xi(s)-(s-A)^{-1}B\omega(s)$, holds and thus

$$D(s-A)^{-1}x = D\xi(s)-D(s-A)^{-1}B\omega(s) = 0-0 = 0.$$

Suppose $D(s-A)^{-1}x=0$, then $x=(s-A)(s-A)^{-1}x-B0$ is a (ξ,ω) representation with $\xi(s)=(s-A)^{-1}x$ is in $Ker\,D$. So $Im\,B\subset V_\Sigma(Ker\,D)$, which implies that $V_\Sigma(Ker\,D)$ is closed by lemma III.7.

□

Let us remark that the condition a) in lemma III.9 is weaker than the condition that the transfer function $D(s-A)^{-1}B$ has only finitely many zeros in plus infinity. This holds even if the input and output space are one dimensional, as can been seen from the next example:

Let X be a Hilbert space and A a generator of a C_0-semigroup on X. Assume furthermore that $b,d\in X$ with $d\notin D(A^*)$, (the domain of the adjoint of A), and $<d,b>_X\neq 0$. Then $<d,A.>_X$ does not have a bounded extension from X to C, but the multiplicity of the zeros in plus infinity of the transfer function $<d,(s-A)^{-1}b>$ is one.

Using the characterization of $V_\Sigma(Ker\,D)$ given in lemma III.5 and theorem III.6 we derive the following equivalent statements for the solvability of DDP.

Theorem III.10.

Assume that $V_\Sigma(Ker\,D)$ is closed, then the following statements are equivalent.

a) DDP is solvable.

b) $Im\,E\subset V_\Sigma(Ker\,D)$.

c) For every q in Q there exists a $\omega(.)$ in $U_{-1}(s)$ (see definition II.12), such that

(3.13) $D(s-A)^{-1}Eq = -D(s-A)^{-1}B\omega(s)$

d) There exists a U(s) in $\left[L(Q,U)\right]_{-1}(s)$ such that

(3.14) $D(s-A)^{-1}BU(s) = D(s-A)^{-1}E$

e)   There exists a U(s) in $\left[\mathcal{L}(\mathcal{Q},\mathcal{U})\right]_{-1}$ (s) and X(s) in $\left[\mathcal{L}(\mathcal{Q},\mathcal{X})\right]_{-1}$ (s) such that

(3.15)
$$\begin{bmatrix} s-A & B \\ D & 0 \end{bmatrix} \begin{bmatrix} X(s) \\ -U(s) \end{bmatrix} = \begin{bmatrix} E \\ 0 \end{bmatrix}.$$

Proof:

This follows trivially from theorems III.3 and III.6, and definitions III.4 and II.13.

$\square$

Remark:

Theorem III.10 is the same as theorem 3.4 and 3.6 in Hautus [19] for the finite dimensional case.

We see from this theorem that under an extra condition on the system the solvability of DDP is equivalent to the solvability of a meromorphic matrix equation. One can easily show that the solvability of (3.14) is always a necessary condition for the solvability of DDP. However it can be seen from example E.10 in appendix E that it is not sufficient; of course in this case $V_\Sigma(Ker\ D)$ will not be closed.

Section III.2: DDP in Time–Domain

In this section we shall state similar results as in the previous section, however now we shall work in the time–domain. Analogous to $V_\Sigma(K)$ we can define the largest open loop invariant subspace.

Definition III.11: $V_{ol}(K)$

Let $V_{ol}(K)$ be the subset of $\mathcal{X}$ which contains all $x_0 \in \mathcal{X}$ with the property that there exists a continuous input $u(.)$ such that the mild solution $x(.)$ of (2.1) i.e. $x(t) = T_A(t)x_0 + \int_0^t T_A(t-s)Bu(s)ds$ is in K for all $t \geq 0$.

Remark:

If K is the kernel of a bounded operator D, then this subspace contains all initial values x_0 such that for some continuous $u(.)$ the output $z(.) = 0$, where z is the output of the system $\dot{x}(t) = Ax(t) + Bu(t)$; $x(0) = x_0$; $z(t) = Dx(t)$.

For this subspace we shall give similar results as for $V_\Sigma(K)$ in section III.1. Since the proofs of these results are very similar we shall omit them.

Theorem III.12.

Let $V_{ol}(K)$ be the subspace defined by III.11, then it has the following properties.

a) $V_{ol}(K)$ is the largest open loop invariant subspace in K.

b) If $V_{ol}(K)$ is closed, then $V^*(K)$ exists and is equal to $V_{ol}(K)$.

c) Let $B^0(V_{ol}(K))$ denote the largest subspace of $Im\, B$ that has zero intersection with $V_{ol}(K)$ and let $\{b_1,..,b_{m_0}\}$ be a basis of $B^0(V_{ol}(K))$. Then $V_{ol}(K)$ is closed if and only if there exist functionals $\{f_i\}_{i=1}^{m_0}$ in $D(A')$ such that $<f_i, b_j> = \delta_{ij}$ and $f_i|_{V_\Sigma(K)} = 0$.

d) Let $B^1(V)$ denote $Im\, B \cap V$. Then $V_{ol}(K)$ is closed if and only if $(V_{ol}(K))^\perp \cap D(A')$ is weak * dense in $(V_{ol}(K))^\perp$ and $B^1(V_{ol}(K)) = B^1(\overline{V_{ol}(K)})$

e) If A is a bounded operator, then $V^*(K)$ exists and it is equal to $V_{ol}(K)$.

Remark:

As in the frequency domain it is possible that $V^*(K)$ exists, but $V_{ol}(K)$ is not closed, see example E.12.

Remark:

As an easy corollary of theorem III.12.d) one has the following result. If V is open loop invariant, then $\overline{V}$ is controlled invariant provided that $V^\perp \cap D(A')$ is weak * dense in $V^\perp$ and $B^1(V)=B^1(\overline{V})$.

Lemma III.13.

Let $Ker\, D$ be the kernel of a bounded linear operator D. Then $V_{ol}(Ker\, D)$ is closed if either of the following conditions holds.

a) There exists a q in $\mathbb{N} \cup \{0\}$ such that DA^i has a bounded extension from X to Z (denoted by $\overline{DA^i}$) for $0 \le i \le q+1$, $\overline{DA^i}B = 0$ for $0 \le i < q$ and $\overline{DA^q}Bu \ne 0$ for all $u \ne 0$ in U.

b) $D(s-A)^{-1}B = 0$ for all s in $[\hat{s}, \infty)$ for some $\hat{s}$ in $\mathbb{R}$.

So we see that $V_\Sigma(K)$ and $V_{ol}(K)$ have the same properties, this together with the definition of these subspaces suggests that $V_\Sigma(K)$ and $V_{ol}(K)$ are equal. However this is presently still unknown. It is even an open problem whether $V_\Sigma(K)$ is closed if and only if $V_{ol}(K)$ is closed.

Section III.3: Properties of Controlled Invariant Subspaces

From the previous sections we obtain very nice results concerning controlled invariant subspaces.

We make the following important remark. If V is a closed linear subspace that is controlled invariant, then $V_\Sigma(V) = V_{ol}(V) = V$. This is obvious from lemma III.5, theorem III.6 and theorem III.12.a) and b).

In finite dimensions use is made of the fact that controlled invariance is closed under subspace addition. Unfortunately this no longer holds in infinite dimensions in general. The next lemma gives sufficient conditions for this to hold.

Lemma III.14.

Let V_1 and V_2 be closed, controlled invariant subspaces of X. Then $\overline{V_1 + V_2}$ is controlled invariant provided that one of the following conditions holds.

a) $V_1 + V_2$ is closed, so in particular V_1 or V_2 is finite dimensional.

b) There exist functionals $\{f_j\}_{j=1}^{m_0}$ in $D(A')$ such that $f_j|_{V_1 + V_2} = 0$ and $\langle f_j, b_i \rangle = \delta_{ij}$, where b_i is a basis for $B^0(V_1 + V_2)$.

c) $(V_1 + V_2)^\perp \cap D(A')$ is weak $*$ dense in $(V_1 + V_2)^\perp$ and $B^1(V_1 + V_2) = B^1(\overline{V_1 + V_2})$.

Proof:

a) If V_1 and V_2 are controlled invariant, then they are by definition open loop invariant. It is not hard to prove that the sum of these two subspaces is also open loop invariant. If one of these subspaces is finite dimensional, then the sum is a closed subspace, and by theorem II.27 we may conclude that this sum is controlled invariant

b) Consider $V_\Sigma(\overline{V_1 + V_2})$, then by lemma III.5.b) $V_1 \subset V_\Sigma(V_1 + V_2)$ and $V_2 \subset V_\Sigma(V_1 + V_2)$. So by the linearity of $V_\Sigma(V_1 + V_2)$, $V_1 + V_2$ is contained in $V_\Sigma(\overline{V_1 + V_2})$. By lemma III.5.c) $V_\Sigma(\overline{V_1 + V_2}) \subset \overline{V_1 + V_2}$, thus $V_\Sigma(\overline{V_1 + V_2}) = \overline{V_1 + V_2}$.

Let $\{b_1,..,b_{m_1}\}$ be a basis for $\overline{B^0(V_\Sigma(V_1+V_2))}$ then since V_1+V_2 is contained

in $V_\Sigma(V_1+V_2)$ we can extend this basis to a basis of $B^0(V_1+V_2)$. Denote this

basis by $\{b_1,..b_{m_1},...,b_{m_0}\}$. By assumption we have that there exist

functionals $\{f_i\}\subset D(A')$ such that $<f_i,b_j>=\delta_{ij}$ and $f_i|_{\overline{V_1+V_2}}=f_i|_{\overline{V_1+V_2}}=$

$f_i|_{\overline{V_\Sigma(V_1+V_2)}}=0$. From lemma III.7.$iii$) we have that $V_\Sigma(V_1+V_2)$ is a closed

subspace, and thus $V_\Sigma(V_1+V_2)=\overline{V_1+V_2}$ is controlled invariant.

c) The proof of c) is similar to that of b) and therefore it will be
omitted.

$\square$

Remark:

It will be shown in example E.15 that there exist controlled invariant
subspaces whose sum is no longer controlled invariant, contrary to the
finite–dimensional case.

We conclude with a generalization of lemma III.14 and give conditions
under which the union of an infinite nest of controlled invariant subspaces
will be controlled invariant.

Lemma III.15.

a)   Let V_n; $n\geq 1$ be a nest of closed, linear and controlled invariant

   subspaces. Then $V:=\overline{\bigcup_{n\geq 1} V_n}$ is controlled invariant if there exist

functionals $\{f_j\}_{j=1}^{m_0}$ in $D(A')$ such that $f_j|_V=0$ and $<f_j,b_i>=\delta_{ij}$, where b_i is
a basis for $B^0(\overline{\bigcup_{n\geq 1} V_n})$.

b)   Consider the same nest V_n. Let B be one dimensional and suppose

   there exists a closed subspace K such that $V_n\subset K$ and the controllability

subspace, $<T_A(t)|Im\,B>$, is not contained in K. Then $V:=\overline{\bigcup_{n\geq 1} V_n}$ is controlled

invariant if and only if there exists a functional f in $D(A')$ such that
$f|_V=0$ and $<f,b>=1$.

Proof:

a) Let V be $\overline{\bigcup_{n\geq 1} V_n}$, then V_n is closed loop invariant and contained in V for

all n, thus by lemma III.5.b) V_n is contained in $V_\Sigma(V)$. Since $\{V_n\}$ is a

nest, we have $\overline{\bigcup_{n\geq 1} V_n}\subset V_\Sigma(V)$. By definition of $V_\Sigma(V)$ we have that

$$\overline{V_\Sigma(V)} \subset V = \bigcup_{n \geq 1} V_n \ , \quad \bigcup_{n \geq 1} V_n \subset V_\Sigma(V) \subset V. \text{ Thus } \overline{V_\Sigma(V)} = V.$$

Let $\{b_1,..b_{m_1}\}$ be a basis for $B^0(V_\Sigma(V))$ then since $\bigcup_{n \geq 1} V_n$ is contained in $V_\Sigma(V)$ we can extend this basis to a basis of $B^0(\bigcup_{n \geq 1} V_n)$. Denote this basis by $\{b_1,..b_{m_1},..,b_{m_0}\}$. By assumption we have that there exist functionals $\{f_i\} \subset D(A')$ such that $<f_i,b_j> = \delta_{ij}$ and $f_i|_V = f_i|_{V_\Sigma(V)} = 0$. From lemma III.7.iii) we have that $V_\Sigma(V)$ is a closed subspace, and thus $V_\Sigma(V) = V$ is controlled invariant.

$\square$

b)

(if): see a)

(only if): Suppose V is controlled invariant and $b \in V$, then V is also $T_A(t)$-invariant. Furthermore since K is closed we have that $V \subset K$. $<T_A(t)|Im\ B>$ is the smallest $T_A(t)$-invariant subspace containing $Im\ B = span\{b\}$.

So $<T_A(t)|Im\ B> \subset V \subset K.$ ⊶

So $B^0(V) = span\{b\}$, and III.7 gives the desired result

$\square$

In this chapter we have analyzed the problem of the existence of the largest controlled invariant subspace. As we have seen, this subspace does not necessarily exist and when it does exist it is not necessarily the largest open loop or the largest frequency invariant subspace and it is difficult to verify when it exists. In the following chapter we consider a special class of systems for which we can give readily verifiable necessary and sufficient conditions for the existence of the largest controlled invariant subspace.

CHAPTER IV: CONTROLLED INVARIANCE FOR DISCRETE SPECTRAL SYSTEMS

In this chapter we shall consider again the following linear controlled system described by the set of equations

(4.1a) $\qquad \dot{x}(t) = Ax(t) + Bu(t)$

(4.1b) $\qquad z(t) = Dx(t)$

but instead of considering this system in a general Banach space we shall assume that the state space $\mathcal{H}$ is a separable Hilbert space and furthermore the input space $\mathcal{U}$ is assumed to be one dimensional. As in the previous chapter A is a generator of a C_0-semigroup on $\mathcal{H}$, D is a bounded linear operator from $\mathcal{H}$ to the Banach space $\mathcal{Z}$ and since the input space is one dimensional we have $Bu = bu$ with $b \in \mathcal{H}$.

In chapter 2 we introduced the concept of controlled invariance. In this chapter we shall derive for a class of spectral systems a complete characterization of all controlled invariant subspaces of (4.1) contained in the kernel of D. This characterization is given in the terms of the zeros of the transfer function $D(s-A)^{-1}B$ of system (4.1). As a consequence of this we derive necessary and sufficient conditions for the existence of $V^*(Ker\,D)$, the largest controlled invariant subspace in the kernel of D

That there exists a close relationship between controlled invariant subspaces in $Ker\,D$ and zeros is well known in finite dimensions and can be illustrated by the following problem. We can ask a very simple question. What is the form of all one dimensional controlled invariant subspaces in the kernel of D?

Let span$\{v\}$ be such a subspace, then since it is controlled invariant it is also $A + BF$ invariant, for some feedback law F, see definition II.28 and lemma I.3.a). Thus v is an eigenvector of $A + BF$. We shall distinguish between two cases

one: $Fv = 0$

In this case v is an eigenvector of A and in the kernel of D.

two: $Fv \neq 0$

Without loss of generality we may assume that $Fv = 1$. So that $(A + BF)v = \alpha v$ implies $(\alpha I - A)v = b$. If we assume that α is in $\rho(A)$, then $v = (\alpha I - A)^{-1}b$. Since v is in the kernel of D we obtain $D(\alpha I - A)^{-1}b = 0$. Thus α is a zero of the transfer function. So we see that a one–dimensional controlled invariant subspace is either an eigenvector of A in the kernel of D or it is span$\{v\}$ where v satisfies the equations $Dv = 0$ and $(\alpha I - A)v = b$ for some α in $\mathbb{C}$. If this α is an element of the resolvent set of A, then it is a zero of the transfer function.

For finite dimensional systems the concept of zeros is very well understood, see e.g. Davison & Wang [15]. In infinite dimensions however only a few articles have been published, see Pohjolainen [30].

The proof of the existence of $V^*(Ker\,D)$ is based on the result on pole placement from Sun [37]. To get an idea that there exists a relationship between the existence of $V^*(Ker\,D)$ and the problem of pole placement we refer to the finite dimensional case. It is well–known (Wonham [42, p. 112 and 113]) that if the single input system (4.1) is controllable and $D \neq 0$, then $\sigma(A+BF|_{V^*(KerD)})$ is fixed for all F satisfying $(A+BF)V^*(Ker\,D) \subset V^*(Ker\,D)$. In this paper we shall prove a similar result for spectral systems, see theorem IV.5. Otherwise we have from the results of Sun [37] a restriction on all possible sets $\sigma(A+BF)$, for some F. Hence especially on the fixed part of $\sigma(A+BF|_{V^*(KerD)})$, and this will give a condition for the existence of $V^*(Ker\,D)$.

The organization of this chapter will be as follows. In section IV.1 we shall recall some facts and properties of discrete spectral operators and for this class of operators we shall derive a characterization of all $T_A(t)$ invariant subspaces.

Properties of the zeros of the class of spectral systems will be given in section IV.2.

In section IV.3 the main theorems of this chapter will be presented. In this section we shall give a full description of all controlled invariant subspaces in the kernel of D. In particular we shall give necessary and sufficient conditions for the existence of $V^*(Ker\,D)$.

Application of these theorems will be given in section IV.4. The examples in this section were partly calculated by L. Nooitgedagt [26].

We remark here that in order to improve the readability we have restricted our attention to systems which are approximately controllable and initially observable, (see Curtain & Pritchard [9, p. 60 and 69]) for a more general theory we refer the reader to Zwart [47].

Section IV.1: Discrete Spectral Operators

In this section we shall give the definition and some properties of discrete spectral operators. For more detail about these operators we refer the reader to Dunford & Schwartz [18].

Definition IV.1: Discrete Operator

A linear operator A from $\mathcal{H}$ to $\mathcal{H}$ is discrete if there exists a number λ in its resolvent set for which the resolvent $R(\lambda;A) := (\lambda I - A)^{-1}$ is compact.

Lemma IV.2.

If A is discrete, then

a) its spectrum, $\sigma(A)$, is a denumerable set of points with no finite limit point;

b) The resolvent $R(\lambda, A)$ is compact for every λ not in $\sigma(A)$.

c) Every λ_0 in $\sigma(A)$ is a pole of finite order $\theta(\lambda_0)$ of the resolvent and if, for some positive integer k, x satisfies the equation

$$(A - \lambda_0 I)^k x = 0$$

then x satisfies the equation

$$(A - \lambda_0 I)^{\theta(\lambda_0)} x = 0$$

The set of all vectors x satisfying the equation $(A - \lambda_0 I)^{\theta(\lambda_0)} x = 0$ is a finite dimensional linear space, called the space of generalized eigenvectors of A corresponding to the eigenvalue λ_0;

d) If

(4.2) $$P(\lambda_0) = \frac{1}{2\pi i} \int_\Gamma (\lambda I - A)^{-1} d\lambda ,$$

where Γ is a small closed curve surrounding only the eigenvalue λ_0 and Γ is traversed once in the positive sense, then $P(\lambda_0)$ projects $\mathcal{H}$ onto the space of generalized eigenvectors corresponding to λ_0.

Proof:

See Dunford & Schwartz [18, lemma XIX 2.2.]. $\qquad\qquad\qquad\square$

Remark:

The spectrum of A shall be denoted by $\{\lambda_n\}\ n \geq 1$.

Definition IV.3. Discrete Spectral Operator.

A discrete operator is spectral if the spectral projections $P(\lambda_j)$ defined by (4.2) satisfy

a) The family of sums of finite collections of projections $P(\lambda_j)$ is uniformly bounded and

b) No non zero x in $\mathcal{H}$ satisfies all of the equations $P(\lambda_j)x = 0$, λ_j in $\sigma(A)$.

Remark:

The spectral projections $P(\lambda_j)$ are not necessarily selfadjoint.

Lemma IV.4.

If A is a discrete spectral operator, then the spectral projections $\{P(\lambda_j),\ \lambda_j$ in $\sigma(A)\ \}$ generate an uniformly bounded Boolean algebra with the completeness property:

$$(4.3) \qquad \sum_{j=1}^{\infty} P(\lambda_j) = I$$

where the convergence is in the strong topology.

Proof:

See Dunford & Schwartz [18, XVIII.1.].   □

If a subspace V of $\mathbf{R}^n$ is invariant with respect to a diagonal matrix, then V must be of the form span$\{e_i,\ i \in J \subset \{1..n\}\ \}$ where e_i is the i'th basis vector of $\mathbf{R}^n$. For the class of discrete spectral operators a similar theorem holds. First we shall recall a lemma of Dunford and Schwartz [18] that gives some invariance properties of the spectral projections, $P(\lambda_i)$.

Lemma IV.5.

Let A be a discrete spectral operator and λ_j an element of $\sigma(A)$. Then $D(A) \supset P(\lambda_j)\mathcal{H}$, the subspace $P(\lambda_j)\mathcal{H}$ is A-invariant, $AP(\lambda_j)x = P(\lambda_j)Ax$ for all x in $D(A)$ and $\sigma(A|P(\lambda_j)\mathcal{H}) = \{\lambda_j\}$.

Proof:

See Dunford & Schwartz [18, p. 2294].   □

With these lemmas we can now give a complete characterization of all $T_A(t)$–invariant subspaces for the class of discrete spectral generators.

Theorem IV.6.

Let the discrete spectral operator A generate the C_0–semigroup $T_A(t)$, then a closed linear subspace V of $\mathcal{H}$ is $T_A(t)$–invariant if and only if

$$(4.4) \qquad V = \sum_{i=1}^{\infty} W_i$$

where W_i is a subspace of $\mathcal{H}$ which is contained in $P(\lambda_i)\mathcal{H}$ and is A-invariant.

The summation (4.4) is in the strong topology, i.e. for all $x \in V$ there exist $\{w_i; i \in \mathbb{N}$, with $w_i \in W_i\}$ such that $\sum_{i=1}^{n} w_i$ converges to x for $n \to \infty$. On the other hand, if $\{w_i; i \in \mathbb{N}$, with $w_i \in W_i\}$ is such that $\sum_{i=1}^{n} w_i$ converges to x for $n \to \infty$, then the limit is in V.

Furthermore the spectrum of A restricted to V is equal to the set of all $\lambda_i \in \sigma(A)$ such that the corresponding W_i is not the zero subspace.

Proof:

(if):

Since the dimension of $P(\lambda_i)\mathcal{H}$ is finite, the dimension of W_i must also be finite. So W_i is a closed linear subspace of $\mathcal{H}$. Furthermore $P(\lambda_i)\mathcal{H}$ is contained in $D(A)$, and with lemma I.7 we may conclude that W_i is $T_A(t)$-invariant.

Every x in V is the limit of a sequence x_n, with $x_n \in \sum_{i=1}^{n} W_i$. So with the above we have $T_A(t)x_n$ is in $\sum_{i=1}^{n} W_i$. $T_A(t)$ is a bounded linear operator thus $T_A(t)x_n$ converges to an element in $\mathcal{H}$, but also to an element of V since $T_A(t)x_n$ is in V. So V is $T_A(t)$-invariant.

(only if):

Let V be a $T_A(t)$-invariant subspace. Since $\sigma(A) = \{\lambda_i\}$, $i \in \mathbb{N}$ we have that the resolvent set, $\rho(A)$, is connected. With lemma I.4 this implies that V is also $(\lambda I - A)^{-1}$ invariant for *all* λ in the resolvent set of A. So

$$P(\lambda_i)V = \frac{1}{2\pi i} \int_\Gamma (\lambda I - A)^{-1} V \, d\lambda \subset V; \text{ see (4.2)}$$

So $P(\lambda_i)V \subset (P(\lambda_i)\mathcal{H}) \cap V$. Using the fact that $P(\lambda_i)$ is a projection we get

$(P(\lambda_i)\mathcal{H})\cap V \;=\; (P(\lambda_i)\mathcal{H})\cap(P(\lambda_i)V)\subset P(\lambda_i)V.$

Thus $P(\lambda_i)V=(P(\lambda_i)\mathcal{H})\cap V.$

If we set W_i equal to $P(\lambda_i)V$ and $x_i:=P(\lambda_i)x;\; x\in V$, then $A(W_i) \;=\; AP(\lambda_i)V$

$= AP(\lambda_i)P(\lambda_i)V \;=\; AP(\lambda_i)\Big\{P(\lambda_i)\mathcal{H}\cap V\Big\} \;\subset\; AP(\lambda_i)\Big\{V\cap D(A)\Big\}=P(\lambda_i)A\Big\{V\cap D(A)\Big\}$

$\subset P(\lambda_i)V=W_i;$ see lemma IV.5. So $AW_i\subset W_i.$

By (4.3) and definition IV.3 every x in $\mathcal{H}$ can be uniquely written as

$\sum\limits_{i=1}^{\infty} x_i.$ Hence $V=\sum\limits_{i=1}^{\infty} W_i.$

We shall now prove the last assertion.

Let J denote the index set of all $\lambda_i\in\sigma(A)$ such that $\dim(W_i)$ is larger than zero. Since $W_i\subset P(\lambda_i)\mathcal{H}$ is finite dimensional and A-invariant it must contain an eigenvector corresponding to λ_i, thus $\{\lambda_i;\; i\in J\}\subset\sigma(A|_V).$
From lemma I.4 we have that for all $\lambda\in\rho_\infty$, $(\lambda I-A)^{-1}|_V$ is a bounded linear operator from V to V and since A is discrete it is also a compact operator. Furthermore it is the inverse of $(\lambda I-A|_V)$. So $A|_V$ is a discrete operator and hence the spectrum of $A|_V$ is a pure point spectrum. Furthermore we have from corollary I.10 that $\sigma(A|_V)\subset\sigma(A)$. Let λ_0 be an element from $\sigma(A|_V)$, then there exists a v in V such that $A|_V(v)=\lambda_0 v.$ From the above we have that $W_0=P(\lambda_0)V.$ We shall show that $W_0=P(\lambda_0)V$ is non zero, but this is obvious since $P(\lambda_0)v=v.$ So it is shown that $\{\lambda_i;\; i\in J\}=\sigma(A|_V).$ $\qquad\square$

With this theorem we can prove some interesting corollaries that are already known but our proof is much simpler, see Curtain and Pritchard [9, p.61].

Let A be of the following form $A=\sum\limits_{i=1}^{\infty}\lambda_i<.,\phi_i>_{\mathcal{H}}\phi_i,$ with $\{\phi_i\}$ an

orthonormal basis of $\mathcal{H}$, $\lambda_i\in\mathbb{R}$, $\lambda_i\neq\lambda_j$ and $\sum\limits_{i=1}^{\infty}\lambda_i^{-2}<\infty.$ If we set the domain of A equal to all x in $\mathcal{H}$ such that $\sum\limits_{i=1}^{\infty}|\lambda_i<x,\phi_i>_{\mathcal{H}}|^2$ exists, then A is a discrete spectral operator with spectral projections $P(\lambda_i)=<.,\phi_i>\phi_i.$ Furthermore if $\sup\{\lambda_i|i\in\mathbb{N}\}<\infty$, then A generates a C_0-semigroup $T_A(t)$, see Curtain [5] and

$$T_A(t)=e^{At}=\sum\limits_{i=1}^{\infty}e^{\lambda_i t}<.,\phi_i>_{\mathcal{H}}\phi_i$$

Let D be a bounded linear operator from $\mathcal{H}$ to a Hilbert space $\mathcal{Z}$. This operator can be seen as an observation. An important concept in system theory is the nonobservable subspace V_0 i.e. the set of all initial conditions x_0 such that $D T_A(t) x_0 = 0;\ \forall t \ge 0$.

Corollary IV.7.

If A and D satisfy the properties as stated above, then V_0 is $\overline{\text{span}}\{\phi_i | D(\phi_i) = 0\}$. So $V_0 = \{0\}$ if and only if $D(\phi_i) \ne 0$ for all i in $\mathbb{N}$.

Proof:

From Curtain [6] we obtain that V_0 is the largest subspace of $\mathcal{H}$ that is semigroup invariant and in the kernel of D. From theorem IV.6 and the special structure of A we have that $V_0 = \overline{\text{span}}\{\phi_i | i \in J \subset \mathbb{N}; D(\phi_i) = 0\}$. Combining these results for V_0 we conclude this corollary.

□

If the nonobservable subspace equals the zero set, then the pair (D,A) is initially observable, see [9, p. 69]. Dual to the concept of initially observable is the concept of approximately controllable, i.e. the system (A,B) is approximately controllable if the set of all states that can by reached from zero with an arbitrary input is dense in $\mathcal{H}$, or $\{x \in \mathcal{H} | \exists t \ge 0, \exists u(.)$

$$s.t\quad x = \int_0^t T_A(t-s) B u(s) ds\ \}$$ is a dense subset of $\mathcal{H}$. The system (A,B) is approximately controllable if and only if the system (B^*, A^*) is initially observable. For our special system we now have the following corollary.

Corollary IV.8.

If A satisfies the same properties as in corollary IV.7 and b is an element of $\mathcal{H}$, then the system (A,b) is approximately controllable if and only if $<b, \phi_i> \ne 0$ for all i in $\mathbb{N}$.

Proof:

This is the dual of corollary IV.7.

□

As in theorem IV.6 we can pose the question what the $T_{A+BF}(t)$–invariant subspaces look like. This question is in general not solvable even if A is a discrete spectral operator since $A+BF$ need not be a discrete spectral operator. Furthermore we are not interested in all $T_{A+BF}(t)$–invariant subspaces but only in those which are in the kernel of D. Before we can give a complete description of all $T_{A+BF}(t)$–invariant subspace we must investigate the notion of the zeros of a transfer function.

Section IV.2: Zeros and Invariance

In this section we shall discuss the relation between the zeros of the system (4.1) and its controlled invariant subspaces in the kernel of D.

Although we shall only consider systems that are initially observable and approximately controllable we shall define zeros in a more general setting.

Definition IV.9: Zero

An element μ of $\mathbb{C}$ is called a zero for the system (4.1) if there exist nonzero $x \in \mathcal{H}$ and a $u \in \mathcal{U}$ such that

$$\begin{bmatrix} \mu - A & B \\ D & 0 \end{bmatrix} \begin{bmatrix} x \\ u \end{bmatrix} = 0.$$

Remark:

This definition can also be found in Davison and Wang [15].

Remark:

If $\mu \in \rho(A)$, then μ is a zero if and only if $D(\mu I - A)^{-1}B = 0$.

Lemma IV.10.

Let V be a closed subspace that is $T_{A+BF}(t)$–invariant and contained in the kernel of D. Assume further that $\lambda \in \sigma_p(A+BF|_V)$, the point spectrum of $A+BF|_V$, then λ is a zero for the system (D,A,B).

Proof:

By assumption there exists an x in V such that $(A+BF)x = \lambda x$. Now define $u = -Fx$, then $\begin{bmatrix} \lambda - A & B \\ D & 0 \end{bmatrix} \begin{bmatrix} x \\ u \end{bmatrix} = \begin{bmatrix} (\lambda - A)x - BFx \\ Dx \end{bmatrix} = \begin{bmatrix} 0 \\ 0 \end{bmatrix}$. So λ is a zero for the system (D,A,B).

$\square$

Remark:

As can be seen from the proof we have not used the special structure of system (4.1). So the result remains true if our state space is a general Banach space.

In the sequel of this section we shall consider system (4.1) with some additional assumptions and discuss the concept of zeros for this special kind of system. The additional assumptions we make on the system (4.1) are:

(Δ1)   The generator A is a discrete spectral operator with spectral decomposition

$$A = \sum_{i=1}^{\infty} \lambda_i P(\lambda_i)$$

where $\lambda_i \neq \lambda_j$ (for all $i \neq j$) and $\dim P(\lambda_i) = 1$ $(i \geq 1)$. Without loss of generality we may assume that the $P(\lambda_i)$ $(i \geq 1)$ are selfadjoint operators in $\mathcal{H}$, see Wermer [40]. The normalized eigenvector of A corresponding to $P(\lambda_i)$ will be denoted by ϕ_i $(i \geq 1)$,

(Δ2) $b_i := \langle b, \phi_i \rangle_{\mathcal{H}} \neq 0$, for all $i \geq 1$.

(Δ3)   For all $i \in \mathbb{N}$, $D\phi_i \neq 0$.

Let us remark that (Δ2) is the controllability assumption and (Δ3) is the observability assumption, see corollary IV.7 and IV.8.

Lemma IV.11.

Assume that the system (4.1) satisfies conditions (Δ1), (Δ2) and (Δ3). If μ is a zero of the system (4.1), then $\mu \in \rho(A)$ and $D(\mu - A)^{-1}B = 0$.

Proof:

Let μ be a zero, then there exists a $x \in D(A)$, $x \neq 0$ and $u \in \mathcal{U}$ such that

$$\begin{bmatrix} \mu - A & b \\ D & 0 \end{bmatrix} \begin{bmatrix} x \\ u \end{bmatrix} = 0,$$

If u were to be zero, then x would be an eigenvector of A in the kernel of D. However this would imply by assumption (Δ3) that $x = 0$, providing the contradiction. So $u \neq 0$. Assume that $\mu \in \sigma(A)$, then $0 = P(\mu)0 = P(\mu)\{(\mu - A)x + bu\} = P(\mu)(\mu - A)x + P(\mu)bu = (\mu - A)P(\mu)x + P(\mu)bu = 0 + P(\mu)bu$. Since $u \neq 0$ this implies that $P(\mu)b = 0$, but this is not possible by assumption (Δ2). So $\mu \in \rho(A)$ and the remark below definition IV.9 gives the desired result.   □

With this lemma we can define controlled invariant subspaces associated with a zero of the system (4.1).

Definition IV.12: Z_{μ}^{k}

Let the system (4.1) satisfy assumptions (Δ1), (Δ2) and (Δ3).

For a zero μ of this system we shall define the following nest of subspaces,

$$Z_{\mu}^{k} = span\{(\mu - A)^{-n}b; \ 1 \le n \le min(k, order(\mu))\},$$

where $order(\mu)$ is the order of the zero of $D(s-A)^{-1}b$ as a meromorphic function. This order is always finite, see Rudin [31].

Remark:

By the previous lemma we have that $\mu \in \rho(A)$, and thus Z_{μ}^{k} is well defined.

Lemma IV.13.

Z_{μ}^{k} is a controlled invariant subspace in the kernel of D.

Proof:

By the definition of Z_{μ}^{k} this subspace is finite dimensional and contained in $D(A)$. So from lemma III.29 we only have to prove that Z_{μ}^{k} is (A,B)-invariant.

Let x be an element of Z_{μ}^{k}. Then $x = \sum_{i=1}^{n} \gamma_i(\mu - A)^{-i}b$, where $n = min(k, order(\mu))$,

and thus $(\mu - A)x = \gamma_1 b + \sum_{i=1}^{n-1} \gamma_{i+1}(\mu - A)^{-i}b$. So with $z = \sum_{i=1}^{n-1} \gamma_{i+1}(\mu - A)^{-i}b$ we have that

$Ax = \mu x - z - \gamma_1 b$, and $A(Z_{\mu}^{k}) \subset Z_{\mu}^{k} + Im\, B$. $\qquad\square$

Lemma IV.14.

Assume that the system (4.1) satisfies (Δ1), (Δ2) and (Δ3).

Then $Z_{\mu_i}^{\theta_i} \cap \overline{span}\left\{ Z_{\mu_j}^{\theta_j} \right\}_{j \in J / i} = \{0\}$, where J denotes the index set of all zeros

and $\theta_j = order(\mu_j)$.

Proof:

The complete proof is very long and rather technical; here we shall give the proof in the special case that the order of all invariant zeros is one.

So we have to prove that $(\mu I - A)^{-1}b$ is not in $\overline{span}\left\{ (\mu_i I - A)^{-1}b \right\}_{\mu_i \ne \mu}$.

Suppose that this were true then there exists a sequence γ_i^n such that

$\| (\mu I - A)^{-1}b - \sum_{i=1}^{n} \gamma_i^n(\mu_i I - A)^{-1}b \| < 1/n$. This implies that

(4.5) $\| D(\mu I - A)^{-1} \{ (\mu I - A)^{-1} b - \sum_{i=1}^{n} \gamma_i^n (\mu_i I - A)^{-1} b \} \| < \| D(\mu I - A)^{-1} \| / n$

By the resolvent identity we have that

(4.6) $(\mu I - A)^{-1} (\mu_i I - A)^{-1} b = (\mu - \mu_i)^{-1} \{ (\mu_i I - A)^{-1} b - (\mu I - A)^{-1} b \}$

Using this identity we have

(4.7) $(\mu I - A)^{-1} \{ (\mu I - A)^{-1} b - \sum_{i=1}^{n} \gamma_i^n (\mu_i I - A)^{-1} b \} =$

$$= (\mu I - A)^{-2} b - \sum_{i=1}^{n} \gamma_i^n (\mu - \mu_i)^{-1} \{ (\mu_i I - A)^{-1} b - (\mu I - A)^{-1} b \}$$

Notice that since μ and μ_i are invariant zeros we have that $D(\mu I - A)^{-1} b$ and $D(\mu_i I - A)^{-1} b$ are both zero. With this property and (4.7) we can simplify the expression inside the norm signs of (4.5).

(4.8) $D(\mu I - A)^{-1} \{ (\mu I - A)^{-1} b - \sum_{i=1}^{n} \gamma_i^n (\mu_i I - A)^{-1} b \}$ (4.7)

$$= D(\mu I - A)^{-2} b.$$

Combining (4.8) with (4.5) implies that $D(\mu I - A)^{-2} b$ is zero, so μ is a zero of order two. This is in contradiction with the assumptions. So $(\mu I - A)^{-1} b$ is not in $\overline{span} \{ (\mu_i I - A)^{-1} b \}_{\mu_i \neq \mu}$.

$\square$

Section IV.3: Characterization of all Invariant Subspaces for Spectral Systems

In this section we shall give a complete characterization of all controlled invariant subspaces in $Ker\, D$. This is yet not possible for an arbitrary generator, but there is a large class, introduced by Sun [37], where we can give a complete answer. We shall start by defining this class.

We say that the system (4.1) satisfies condition Δ if it satisfies $\Delta 1$, $\Delta 2$ and $\Delta 3$, see section IV.2, and we have the following extra conditions on the eigenvalues of A

$(\Delta 4)$ $\qquad \inf_{i \neq j} |\lambda_i - \lambda_j| = \delta > 0,$

$(\Delta 5)$ $\qquad \sup_{i \in \mathbb{N}} \sum_{\substack{j=1 \\ i \neq j}}^{\infty} \left| \frac{1}{\lambda_i - \lambda_j} \right|^2 < \infty.$

So these conditions imply that the spectrum of A may not have an accumulation point($\Delta 4$), and furthermore there must be sufficiently growth in their successive distances. For instance if $\lambda_i = -\sqrt{i}$, then this sequence satisfies $\Delta 4$, but not $\Delta 5$.

Before we shall state the main results of this chapter we shall give the result of Sun [37].

Theorem IV.15.

Suppose that the system (A,b) satisfies $\Delta 1$, $\Delta 2$, $\Delta 4$ and $\Delta 5$. Then for a given sequence of complex numbers $\Omega = \{\nu_1,..,\nu_n,..\}$, in order that there exist $F \in \mathcal{L}(\mathcal{H},\mathbb{R})$ such that the spectrum of the operator $A + bF$ satisfies

$$\sigma(A+bF) = \sigma_p(A+bF) = \Omega,$$

a necessary and sufficient condition is that there exists a subsequence n_j $(j \in \mathbb{N})$ in $\mathbb{N}$ such that

$$\sum_{j \in \mathbb{N}} \left| \frac{\lambda_{n_j} - \nu_j}{b_{n_j}} \right|^2 < \infty$$

where λ_{n_j} is the n_j'th eigenvalue of A and $b_{n_j} = <\phi_{n_j},b>_{\mathcal{H}}$; ϕ_{n_j} is the eigenvector corresponding to λ_{n_j}.

Proof:

See Sun [37]. $\qquad\qquad\qquad\qquad\qquad\qquad\qquad\qquad\qquad\qquad\qquad\qquad \square$

Our main results are:

Theorem IV.16.

Suppose that (4.1) satisfies conditions Δ. Then a closed linear subspace V in $Ker\, D$ is controlled invariant if and only if there exists a subset of $\mathbb{N}$, denoted by J, such that:

a)

$$(4.9) \qquad V = \sum_{j \in J} Z_{\mu_j}^{k_j}$$

where μ_j are distinct zeros,

b) $\dim(Z_{\mu_j}^{k_j}) > 0$ for all j in J and $\dim(Z_{\mu_j}^{k_j}) = 1$ for all but finitely many j in J and

c) there exists a subsequence n_j $(j \in J)$ in $\mathbb{N}$ such that

$$(4.10) \qquad \sum_{j \in J} \left| \frac{\lambda_{n_j} - \mu_j}{b_{n_j}} \right|^2 < \infty$$

where λ_{n_j} is the n_j'th eigenvalue of A and $b_{n_j} = <\phi_{n_j}, b>_{\mathcal{H}}$; ϕ_{n_j} is the eigenvector corresponding to λ_{n_j}.

We shall postpone the proof of this and the next theorem to the end of this section.

Remark:

From theorem IV.5 and the proof of this theorem we have that $\sigma(A + BF|_V) = \sigma_p(A + BF|_V) = \{\mu_j; j \in J\}$.

So if the subspace V in theorem IV.16 is feedback invariant with respect to a stable semigroup, then $\{\mu_j; j \in J\} \subset \{s \in \mathbb{C}; Re(s) < 0\}$.

With this theorem we can solve the existence of $V^*(Ker\ D)$.

Theorem IV.17.

Suppose that the system (4.1) satisfies condition Δ.

Then $V^*(Ker\ D)$ exists if and only if the following conditions hold:

i) For all but finitely many j's the order of the zero, μ_j, is one and

ii) There exists a subsequence n_j $(j \in J)$ in $\mathbb{N}$ such that

$$(4.11) \qquad \sum_{j \in J} \left| \frac{\lambda_{n_j} - \mu_j}{b_{n_j}} \right|^2 < \infty$$

where J is the index set of *all* invariant zeros and λ_{n_j}, b_{n_j} are the same as in theorem IV.16.

Remark:

For the proof of the sufficient part in both theorems, condition $\Delta 5$ can be omitted, see Clarke and Holland [4].

Combining this theorem with the results proved in chapter 3 gives for a class of systems properties concerning the asymptotic distribution of the zeros, see also the next section.

Corollary IV.18.

Assume that the system (4.1) satisfies Δ and assume further that there exists a constant $q \in \mathbb{N}$ such that DA^i has a bounded extension from $\mathcal{H}$ to $\mathcal{Z}$ for $0 \le i \le q+1$, $DA^i b = 0$ for $0 \le i < q$ and $DA^q b \ne 0$. Then the zeros $\{\mu_j\}$ of this system satisfy

i)   For all but finitely many j's the order of the zero, μ_j, is one and

ii) There exists a subsequence n_j $(j \in J)$ in $\mathbb{N}$ such that

$$\sum_{j \in J} \left| \frac{\lambda_{n_j} - \mu_j}{b_{n_j}} \right|^2 < \infty$$

where J is the index set of *all* invariant zeros and λ_{n_j}, b_{n_j} are the same as in theorem IV.16.

Proof:

This is a direct consequence of lemma III.9 and IV.17. $\square$

Before we can prove theorem IV.16 and IV.17 we must first prove some lemmas.

Lemma IV.19.

If A satisfies conditions $\Delta 1$, $\Delta 4$ and $\Delta 5$, then $A+BF$ is a discrete spectral operator for all bounded feedback laws F. The spectral projections of $A+BF$ will be denoted by $P_F(\nu_i)$. Furthermore if i is sufficiently large, then $\dim(P_F(\nu_i)\mathcal{H}) = 1$.

Proof:

See Sun [37] theorem 2.1 and page 734. See Kato [22, p. 196] for the fact that $(\lambda I - A - BF)^{-1}$ is a compact operator. $\square$

Lemma IV.20.

Suppose that the system (4.1) satisfies assumption Δ.

If W is an subspace of $P_F(\mu)\mathcal{H}$ contained in the kernel of D which is $(A+BF)$-invariant with $\dim(W) = k$, $k > 0$, then μ is a zero with $k \leq \text{order}(\mu)$ and $W = Z_\mu^k$.

Proof:

Since W is a finite dimensional, $(A+BF)$-invariant subspace in $P_F(\mu)\mathcal{H}$ there exists a non zero vector e_1 in W such that $(A+BF - \mu I)e_1 = 0$. Thus

(4.12) $\qquad (\mu - A)e_1 - bFe_1 = 0 \ \text{ and } \ De_1 \subset D(W) = 0.$

So μ is a zero and thus by lemma IV.11 an element of $\rho(A)$. So $e_1 = (\mu - A)^{-1}bFe_1$. Since $e_1 \neq 0$ we have that $Fe_1 \neq 0$, and so $\text{span}\{e_1\} = Z_\mu^1$. Now all eigenvectors of $A+BF|_W$ will satisfy $e_1 = (\mu - A)^{-1}bFe_1$ and so $A+BF|_W$ has only one eigenvector.

Since W is finite dimensional, $(A+BF)W \subset W$, $\sigma(A+BF|_W) = \{\mu\}$ and $A+BF|_W$ has only one eigenvector we have that $W = \text{span}\{e_1,..,e_k\}$ where e_i satisfies: $(A+BF - \mu)e_1 = 0$ and $(A+BF - \mu)e_i = e_{i-1}$, $2 \leq i \leq k$.

This last equation implies that $e_i = (\mu - A)^{-1}bFe_i - (\mu - A)^{-1}e_{i-1}$, $2 \leq i \leq k$. Thus by induction it is now easy to prove that $i \leq \text{order}(\mu)$ and $\text{span}\{e_1,..,e_i\} \subset Z_\mu^i$, $2 \leq i \leq k$. Equality follows easily from the fact that the dimensions are equal.

$\qquad\qquad\qquad\qquad\qquad\qquad\qquad\qquad\qquad\qquad\qquad\qquad\qquad\qquad\qquad$ □

Proof of theorem IV.16.

(only if)

Let V be $T_{A+BF}(t)$-invariant, then by lemma IV.19 and theorem IV.6 V is of the following form

(4.13) $\qquad V = \sum_{i=1}^{\infty} W_i$

where W_i is a $(A+BF)$-invariant subspace of $P_F(\mu)\mathcal{H}$.

Define J to be the set of indices with W_i not the zero subspace. By lemma IV.19, we have for all but finitely i in J, $\dim(W_i) = \dim(P_F(\mu_i)\mathcal{H}) = 1$. Since V is in $\text{Ker } D$ we have that W_i is in $\text{Ker } D$, so by lemma IV.19 we have that $W_i = Z_{\mu_i}^{k_i}$, $i \in J$.

Since $\{\mu_i; i \in J\}$ is contained in the spectrum of $A+BF$ we have from Sun [37] that (4.10) must hold.

(if)

Let J_1 be the subset of J that contains all indices j such that $\dim(Z_{\mu_j}^{k_j})$ is one, then by definition IV.12 $Z_{\mu_j}^{k_j} = Z_{\mu_j}^1$. By condition $\triangle$ and (4.10) we may apply theorem IV.15. So there exists a bounded feedback law F_1 such that $\sigma(A+BF_1) \supset \{\mu_j \mid j \in J_1\}$.

We shall prove that $V_1 := \overline{\text{span}}\{Z_{\mu_j}^1\}$ is closed loop invariant with respect
$\quad\quad\quad\quad\quad\quad\quad\quad\quad\quad j \in J_1$
to the operator $T_{A+BF_1}(t)$. Let e_j be an eigenvector of $A+BF_1$ with eigenvalue μ_j. We shall prove that $e_j \in \text{Ker } D$. Since e_j is an eigenvector we have that $(A+BF_1)e_j = \mu_j e_j$ or equivalently

(4.14) $\quad\quad\quad (\mu_j I - A)e_j = bF_1 e_j$

Since μ_j is a zero we have that $\mu_j \in \rho(A)$ and so we can premultiply (4.14) with $(\mu_j I - A)^{-1}$. So

(4.15) $\quad\quad\quad e_j = (\mu_j I - A)^{-1} bF_1 e_j$

This equation implies that $De_j = D(\mu_j I - A)^{-1} bF_1 e_j$. This last expression is zero since μ_j is a zero. So $e_j \in \text{Ker } D$, and we can conclude from lemma IV.19 that $\text{span}\{e_j\} = Z_{\mu_j}^1$.

So for every j in J_1 we have that $Z_{\mu_j}^1$ is $T_{A+BF_1}(t)$–invariant. Since the feedback law F_1 is independent of μ_j, $j \in J_1$, we have that $V_1 = \overline{\text{span}}\{Z_{\mu_j}^1\}$ is
$\quad\quad\quad\quad\quad\quad\quad\quad\quad\quad\quad\quad\quad\quad\quad\quad\quad\quad\quad\quad j \in J_1$
also $T_{A+BF_1}(t)$–invariant.

Let J_2 be the subset of J that contains all indices j such that $\dim(Z_{\mu_j}^{k_j})$ is larger than one. Then by the condition in theorem IV.16 we have that J_2 is finite. Now with lemma IV.13 and III.14.a) we conclude that $V_2 = \text{span}\{Z_{\mu_j}^{k_j}\}$ is controlled invariant.
$\quad\quad\quad\quad j \in J_2$
Since $V = V_1 + V_2$ and V_2 is finite dimensional we can apply lemma III.14.a) to conclude the proof. $\quad\quad\quad\quad\quad\quad\quad\quad\quad\quad\quad\quad\quad\quad\quad\quad\quad\quad \square$

Proof of theorem IV.17.

(if)

Define V to be the closure of the span over all $Z_{\mu_j}^{\theta_j}$, $j \in J$, where μ_j are the invariant zeros and θ_j is their order. From theorem IV.16 we have that this subspace is bounded closed loop invariant and by definition it is the largest closed subspace with this property.

(only if)

By heorem IV.16 we have that $V^*(Ker\,D)$ is of the form $\sum\limits_{j\,\epsilon\,J} Z^{k_j}_{\mu_j}$. If there exists a μ_{j_0} such that j_0 is not in J or k_{j_0} is smaller than θ_{j_0}; the order of μ_{j_0}, then by lemma IV.14 ($\sum\limits_{j\,\epsilon\,J} Z^{k_j}_{\mu_j}+Z^{k_{j_0}}_{\mu_{j_0}}$) is larger than $\sum\limits_{j\,\epsilon\,J} Z^{k_j}_{\mu_j}$ and since $Z^{k_{j_0}}_{\mu_{j_0}}$ is finite dimensional this subspace is by lemma III.14.a) controlled invariant. This is in contradiction to the fact that $V^*(Ker\,D) = \sum\limits_{j\,\epsilon\,J} Z^{k_j}_{\mu_j}$ is the largest closed subspace with this property. So $V^*(Ker\,D) = \sum\limits_{j\,\epsilon\,J} Z^{\theta_j}_{\mu_j}$, where the summation is over all invariant zeros and θ_j = order(μ_j). Conditions i) and ii) are direct consequences of theorem IV.16.

$\square$

Section IV.4: Examples

In this section we shall discuss some examples.

The example that will be discussed in this section is the heated rod with various controls and observations. This can be schematized as below

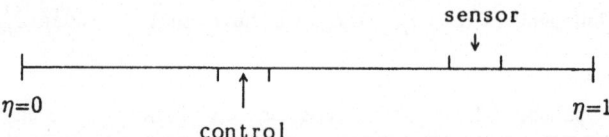

and for the mathematical model we take

(4.16)

$$
\begin{cases}
\dfrac{\partial x}{\partial t} = \dfrac{\partial^2 x}{\partial \eta^2} + b(\eta)u(t) \\[2mm]
x(0,t) = 0 = x(1,t) \\[2mm]
z(t) = \displaystyle\int_0^1 d(\eta)x(t,\eta)d\eta
\end{cases}
$$

This can be formulated as a system of the form (4.1), where we take $\mathcal{H} = L^2(0,1)$ and the system operator A is given by

$$A = \frac{\partial^2}{\partial \eta^2} \quad ; \ D(A) = \{ \ h \in \mathcal{H} : h'' \in \mathcal{H} \ and \ h(0) = 0 = h(1) \ \}.$$

A is self adjoint and has eigenvalues $\{ \ -n^2\pi^2; \ n = 1,..,\infty \ \}$ and eigenvectors $\{ \phi_j(\eta) = \sqrt{2} \cdot sin(j\pi\eta); \ j = 1,..,\infty \ \}$.

In this section we take $b(\eta) = 1_{[0,2/\pi]}(\eta)$, the characteristic function of the interval $[0,2/\pi]$, and we shall investigate the existence of $V^*(Ker\ D)$ for different measurement functions $d(\eta)$.

It is easy to see that A satisfies $\Delta 1$, $\Delta 4$ and $\Delta 5$. Furthermore since

$$(4.17) \qquad b_j = \ <\phi_j, b> \ = \ \int_0^1 1_{[0,2/\pi]}(\eta) \ \sqrt{2} \ sin(j\pi\eta)d\eta \ = \ \frac{-\sqrt{2}}{j\pi} \{ \ cos(2j) - 1 \ \} \neq 0,$$

(A,B) satisfies $\Delta 2$.

With this A and B we shall investigate the existence of $V^*(Ker\ D)$ for three different measurement operators.

Example IV.21.

We shall consider system (4.16) with control operator $b(.) = 1_{[0,2/\pi]}(.)$. In this example we take as measurement operator, $D_1 = \ <.,d_1>$ where

$$(4.18) \qquad\qquad d_1 = 1_{[3/\pi,1]}$$

Since $D_1\phi_j = \ <\phi_j, d_1> = \int_0^1 1_{[3/\pi,1]}(\eta) \ \sqrt{2} \ sin(j\pi\eta)d\eta \ = \ \frac{-\sqrt{2}}{j\pi} \{ \ (-1)^j - cos(3j) \ \} \neq 0,$

we have that the system (A,B,D) satisfies condition Δ.

Using the same technique as in Nooitgedagt [26] we can calculate the transfer function of this system. The transfer function is given by

$$G_1(s) = \frac{\left(1 - cos((2/\pi)\lambda)\right) \left(1 - cos((1 - 3/\pi)\lambda)\right)}{\lambda^3 sin(\lambda)}, \quad where \ \lambda = -i\sqrt{s} \ and$$

(4.19)

$$G_1(0) = -(1 - 3/\pi)^2 / \pi^2$$

For this system we want to investigate the existence of $V^*(Ker\ D_1)$. From theorem IV.17 we have that the existence is determined by the zeros of $G_1(s)$. The zeros of this transfer functions are easily calculated and the set of zeros is given by:

(4.20) $\{\mu_n\} = \{-n^2\pi^4; \ n\in\mathbb{N} \ and \ -n^2\pi^4*(\ 2/(\pi-3))^2; \ n\in\mathbb{N}\}$

Let the zeros be numbered in decreasing order, thus $\mu_j > \mu_{j+1}$.

From theorem IV.17 we have that the existence of $V^*(Ker\,D)$ is determined by the convergence of the sum $\displaystyle\sum_{j\in\mathbb{N}}\left|\frac{\lambda_{n_j}-\mu_j}{b_{n_j}}\right|^2$, where λ_{n_j} is the subsequence of λ_n as in theorem IV.16.

We shall first investigate the single terms in this sum. By simple numerical calculations we have for $j\in\{1,..,100\}/\{30,45,60,75,90\}$ that

$$\left[\frac{\lambda_n-\mu_j}{b_n}\right]^2 \geq \left[\frac{\lambda_1-\mu_j}{b_1}\right]^2 \quad \forall n\in\mathbb{N} \text{ and for } j=30,45,60,75,90 \text{ this minimum is}$$

achieved for respectively $n=89,133,177,222$ and 266. Let min_j denote the index such that

$$\left[\frac{\lambda_n-\mu_j}{b_n}\right]^2 \geq \left[\frac{\lambda_{min_j}-\mu_j}{b_{min_j}}\right]^2 \quad \forall n\in\mathbb{N}. \text{ We then have that}$$

$$\sum_{j=1}^{100}\left[\frac{\lambda_{n_j}-\mu_j}{b_{n_j}}\right]^2 \text{ is larger than } \sum_{j=1}^{100}\left[\frac{\lambda_{min_j}-\mu_j}{b_{min_j}}\right]^2 \text{ for every subsequence}$$

$\{n_j\}$. In the next table we have listed the partial sums, $S_1(m):=$

$$\sum_{j=1}^{m}\left[\frac{\lambda_{min_j}-\mu_j}{b_{min_j}}\right]^2 .$$

In figure 4.1 the numbers of table 4.1 are plotted and we see that in this example we have (numerical) evidence that $V^*(Ker\,D_1)$ does not exist.

Table 4.1

m	$S_1(m)$	m	$S_1(m)$
1	3.8211e+003	20	2.8479e+009
2	7.5736e+004	30	1.8650e+010
3	4.5039e+005	40	8.0862e+010
4	1.6463e+006	50	2.3042e+011
5	4.5795e+006	60	5.8681e+011
6	1.0677e+007	70	1.2725e+012
7	2.1990e+007	80	2.3695e+012
8	4.1308e+007	90	4.1805e+012
9	7.2273e+007	100	7.2350e+012
10	1.1949e+008		

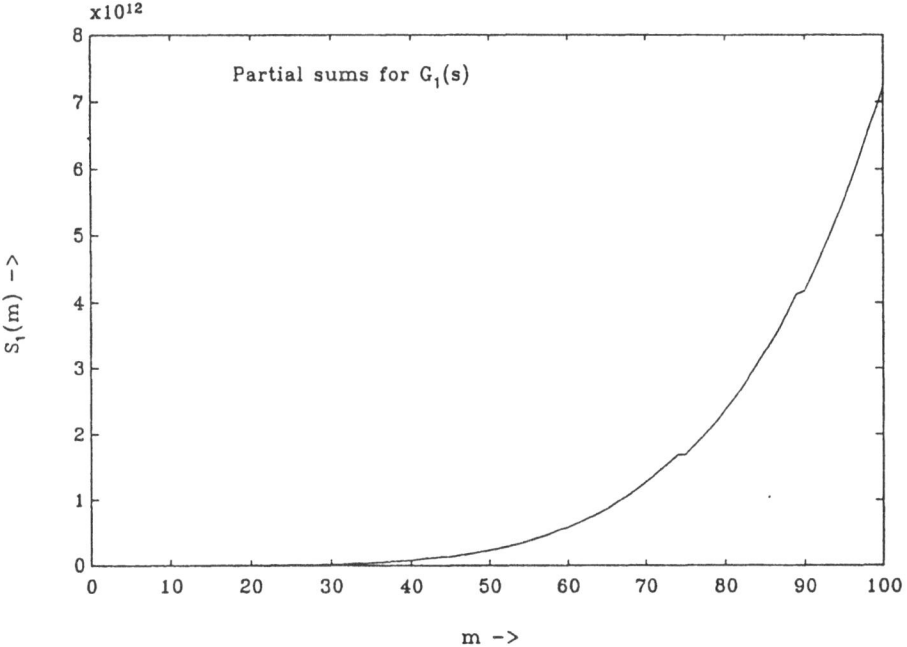

Figure 4.1

For our second example we shall take two measurement functions d_2 and d_3 which are relatively close to each other with $<d_2, b> \neq 0$ and $<d_3, b> \neq 0$, but the

the first is an element of $D(A')$ and the second is not.

Example IV.22.

In this second example we shall investigate the existence of $V^*(Ker\, D)$ for two measurement functions that are close in the $L^2(0,1)$–norm, but, as we shall see in the sequel, $V^*(Ker\, D)$ only exists for one of them. The measurements functions we shall consider are D_2 and D_3, where $D_i = <., d_i>$ with

(4.21) $\qquad d_2(\eta) = \mathbb{1}_{[0,1/\pi]}(\eta)$

and

(4.22) $\qquad d_3(\eta) = \begin{cases} -100\eta^2 + 20\eta & ; \quad 0 \leq \eta \leq 0.1 \\ 1 & ; \quad 0.1 \leq \eta \leq 1/\pi - 0.1 \\ 2000(\eta - 1/\pi)^3 + 300(\eta - 1/\pi)^2; & 1/\pi - 0.1 \leq \eta \leq 1/\pi \\ 0 & ; \quad 1/\pi \leq \eta \leq 1 \end{cases}$

We remark that d_3 is an element of $D(A) = D(A')$. Furthermore we have that $D_3(b)$ is unequal to zero. So from lemma III.9.a) we have directly that $V^*(Ker\, D_3)$ exists. However in this example we shall follow the line of this chapter and we shall give a (numerical) indication for the existence of $V^*(Ker\, D_3)$. Furthermore we shall investigate also the existence of $V^*(Ker\, D_2)$.

Let G_2 and G_3 denote the transfer function of respectively the system (D_2, A, b) and (D_3, A, b). Then

(4.23) $\qquad G_2(s) = \dfrac{(-1/\pi)\lambda * sin(\lambda) - cos(\lambda) + \{1 - cos(\lambda/\pi)\} * cos((1 - 2/\pi) * \lambda)}{\lambda^3 sin(\lambda)}$

where $0 \neq \lambda^2 = -s$, and

(4.24) $\qquad G_2(0) = \dfrac{1}{16}\left\{ \dfrac{1}{\pi^4} + \dfrac{2}{3} * \dfrac{1}{\pi^3} \right\}$

and

(4.25)
$$G_3(s) = \frac{-1000}{\lambda^6 sin(\lambda)} \left\{ 0.001*(1/_\pi - 0.5/_6)\lambda^4 sin(\lambda) + \right.$$

$$0.2*\lambda\left\{ 0.1*\lambda sin(\lambda) + cos(\lambda) - cos(0.9*\lambda) + 3cos((1 - 1/_\pi)*\lambda) + 3cos(1.1 - 1/_\pi)*\lambda) + \right.$$

$$+ cos((1 - 2/_\pi)*\lambda)\left\{ cos(0.1*\lambda) - 1 - 3cos(1/_\pi*\lambda) - 3cos((0.1 - 1/_\pi)*\lambda) \right\} \right\} +$$

$$+ 12\left\{ sin((1 - 1/_\pi)*\lambda) - sin((1.1 - 1/_\pi)*\lambda) + \right.$$

$$\left. + cos((1 - 2/_\pi)*\lambda)*\left[sin(\lambda/_\pi) - sin((1/_\pi - 0.1)*\lambda) \right] \right\} \right\} \quad ; \quad s \neq 0.$$

where $0 \neq \lambda^2 = -s$, and

(2.26)
$$G_3(0) = 1/\pi^4 - 0.388*1/\pi^3 - \frac{359}{3\,000}*1/\pi^2 - \frac{71}{6000}*1/\pi - 0.00005.$$

For both measurement functions we have that $D_i\phi_j \neq 0$, for D_2 this is an easy check, for D_3 however we had to make a numerical check. This we can only do on a finite number of eigenfunctions, and for this measurement we have chosen the first 1000.

As in the previous example we have calculated the zeros of these systems, these zeros are given in table 4.2. The last column of this table contains the eigenvalue of A which is closest to the zero.

Table 4.2

zero-number	G_2	G_3	pole
1	−2.5213e+001	−2.4597e+001	−3.9478e+001
2	−8.8340e+001	−8.8222e+001	−8.8826e+001
3	−1.4581e+002	−1.4010e+002	−1.5791e+002
4	−2.3932e+002	−2.3182e+002	−2.4674e+002
5	−3.5527e+002	−3.5488e+002	−3.5531e+002
6	−4.8216e+002	−4.8464e+002	−4.8361e+002
7	−6.1713e+002	−6.3462e+002	−6.3165e+002
8	−7.9544e+002	−7.9865e+002	−7.9944e+002
9	−9.8002e+002	−9.8268e+002	−9.8696e+002
10	−1.1821e+003	−1.1774e+003	−1.1942e+003
20	−4.3389e+003	−4.3537e+003	−4.3525e+003
30	−9.4845e+003	−9.4845e+003	−9.4847e+003
40	−1.6590e+004	−1.6591e+004	−1.6591e+004
50	−2.5669e+004	−2.5671e+004	−2.5671e+004
60	−3.6710e+004	−3.6725e+004	−3.6725e+004
70	−4.9738e+004	−4.9753e+004	−4.9753e+004
80	−6.4753e+004	−6.4754e+004	−6.4754e+004
90	−8.1730e+004	−8.1730e+004	−8.1730e+004
100	−1.0068e+005	−1.0068e+005	−1.0068e+005

Let $\mu_{2,i}$ and $\mu_{3,i}$ denote the i'th zero of respectively $G_2(.)$ and $G_3(.)$. For the calculated zeros, $\mu_{2,i}$; $2 \leq i \leq 100$, it can be shown that

$$\min_{j \in \mathbb{N}} \left\{ \left[\frac{\lambda_j - \mu_{2,i}}{b_j} \right] \right\} = \left[\frac{\lambda_{i+1} - \mu_{2,i}}{b_{i+1}} \right].$$

So if $S_2(m)$ denotes the partial sum $S_2(m) := \sum_{i=1}^{m} \left[\frac{\lambda_{i+1} - \mu_{2,i}}{b_{i+1}} \right]^2$, then

for $m \leq 100$ we have that $S_2(m)$ is smaller than $\sum_{i=1}^{m} \left[\frac{\lambda_{j_i} - \mu_{2,i}}{b_{j_i}} \right]^2$ for every

subsequence $\{j_i; 1 \leq i \leq 100 \}$ in $\mathbb{N}$.

Calculating this partial sum and the partial sum for $G_3(.)$,

$$S_3(m) := \sum_{i=1}^{m} \left[\frac{\lambda_{i+1} - \mu_{3,i}}{b_{i+1}} \right]^2 ; \quad \text{where} \quad \lambda_{i+1} = -(i+1)^2\pi^2 \quad \text{is the eigenvalue closest}$$

to $\mu_{3,i}$, yields:

Table 4.3.

m	$S_2(m)$	$S_3(m)$
1	5.7929e+002	1.5987e+003
2	7.1984e+003	1.1842e+004
3	1.6011e+004	3.0937e+004
4	1.8017e+004	3.9057e+004
5	1.8027e+004	4.0401e+004
6	1.8710e+004	4.0744e+004
7	3.6107e+004	4.1469e+004
8	9.1562e+004	4.3636e+004
9	1.5939e+005	6.9395e+004
10	1.8121e+005	1.1164e+005
20	1.2466e+006	1.9698e+005
30	2.9993e+006	2.1765e+005
40	8.1812e+006	2.3134e+005
50	1.3083e+007	2.3634e+005
60	2.4109e+007	2.4241e+005
70	3.7819e+007	2.4474e+005
80	5.3018e+007	2.4819e+005
90	8.0695e+007	2.4969e+005
100	9.8908e+007	2.5212e+005

The plots of these partial sums as function of m are given in figure 4.2 and 4.3.

Thus concerning theorem IV.17 we have (numerical) evidence that for $d_3(\eta)$ $V^*(Ker\,D)$ exists, but not for $d_2(\eta)$.

In this example we see that, in spite of the fact that $d_2(\eta)$ and $d_3(\eta)$ are close in L^2–norm, $V^*(Ker\,D)$ exists for d_3 but not for d_2, and this depends on their smoothness.

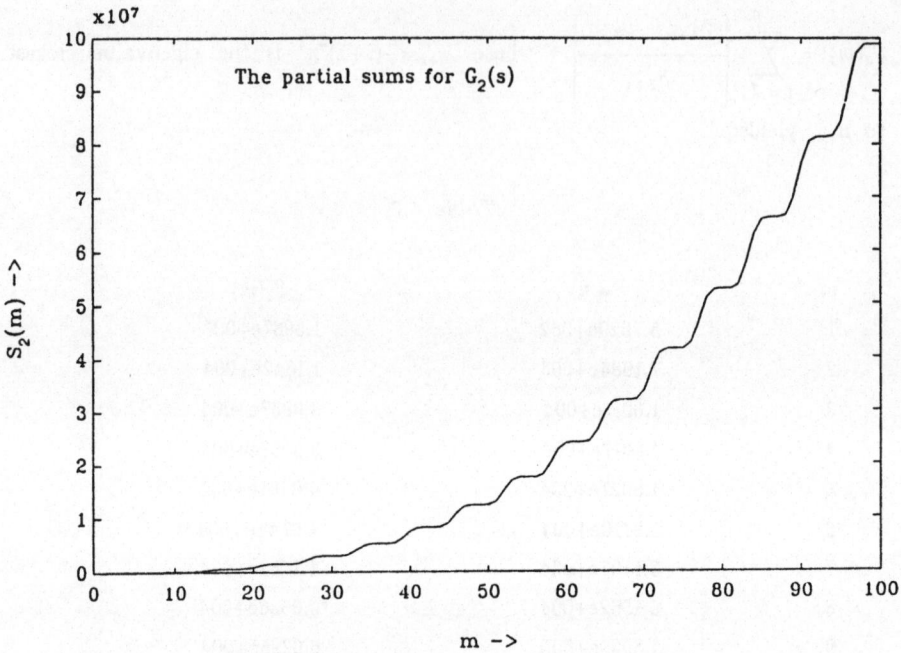

Figure 4.2

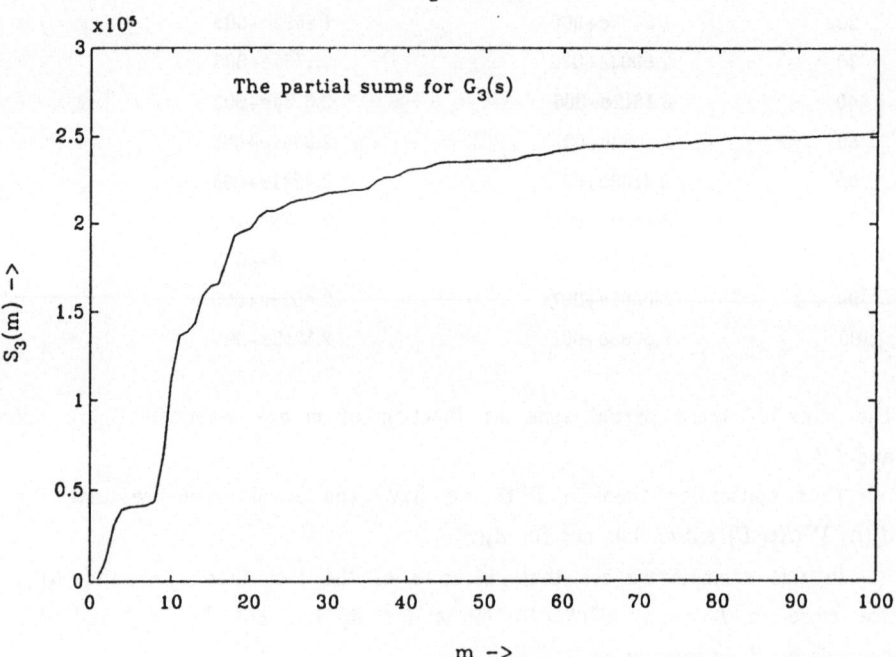

Figure 4.3

CHAPTER V: THE DISTURBANCE DECOUPLING PROBLEM WITH MEASUREMENT FEEDBACK

In this chapter we shall investigate the disturbance decoupling problem, as in chapter III, but instead of using full state feedback we shall now use a compensator.

In Curtain [7] the *Disturbance Decoupling Problem* with *Measurement feedback* (DDPM) was investigated for infinite dimensional systems. Under strong assumptions she was able to prove sufficient conditions for DDPM to be solvable. By examining the proofs given in Curtain [7] more deeply and using the results of the previous chapters, we can obtain weaker conditions for the solvability of DDPM in infinite dimensions. Furthermore we shall give necessary and sufficient conditions such that the compensator is finite dimensional. Instead of working in a general Banach space, as we did for the Disturbance Decoupling Problem, we shall here use a Hilbert space. This condition will simplify the proofs, and it will improve the readability.

We consider the following system Σ on a Hilbert space $\mathcal{H}$:

$$(5.1) \qquad \Sigma: \quad \begin{aligned} \dot{x} &= Ax + Bu \\ y &= Cx \end{aligned}$$

where A is the infinitesimal generator of a strongly continuous semigroup $T_A(t)$ on $\mathcal{H}$, $B \in \mathcal{L}(\mathbb{R}^m, \mathcal{H})$ and $C \in \mathcal{L}(\mathcal{H}, \mathbb{R}^k)$. The following lemma will prove useful in the sequel.

Lemma V.1.

Suppose that V is invariant under the semigroup $T_{A+Q}(t)$, generated by $A+Q$ for $Q \in \mathcal{L}(\mathcal{H})$, and let $Q_0 \in \mathcal{L}(\mathcal{H})$ be such that

$$(Q - Q_0)(V \cap D(A)) = 0$$

Then $T_{A+Q}(t)x = T_{A+Q_0}(t)x$ for all x in V.

Proof:

The generator of $T_{A+Q}(t)|_V$ is $(A+Q)|_{V \cap D(A)}$ and on $V \cap D(A)$ we have that $A + Q = A + Q_0$.

From lemma II.2. we have that $A+Q_0$ generates a semigroup on V. Since the generators of both semigroups are the same on V, the semigroups must be the same. So $T_{A+Q}(t)x = T_{A+Q_0}(t)x \ \forall x \in V$.

$\square$

Section V.1: Conditioned Invariance and (C,A,B)–pairs

The concept of controlled invariance was a property of the system operator A and the input operator B. The dual concept (see lemma V.3) is a property of the system operator and the output operator C. This concept is called conditioned invariance.

Definition V.2: Conditioned Invariance

A *closed* subspace S of $\mathcal{H}$ is conditioned invariant if there exists a bounded operator G such that $T_{A+GC}(t)S \subset S$.

As stated above, this is the dual concept of controlled invariance. The next lemma will give the precise meaning of this dualism.

Lemma V.3.

A closed subspace V of $\mathcal{H}$ is controlled invariant for the system (A,B) if and only if $V^\perp$ is conditioned invariant for the system (B^*,A^*), where Q^* denotes the adjoint operator of Q.

Proof:

This follows trivially from the fact that $T_{A+BF}(t)V \subset V$ if and only if $T^*_{A+BF}(t)V^\perp \subset V^\perp$ and $T^*_{A+BF}(t) = T_{A^*+F^*B^*}(t)$.

$\square$

As is to be expected, the concept of condioned invariance has also a dual open loop version and a frequency domain version.

Theorem V.4.

Let $S \subset \mathcal{H}$ be a closed linear subspace. Then the following assertions are equivalent.

i) S is a conditioned invariant subspace.

ii)   There exists a generator Q of a C_0–semigroup on $S^\perp$, $T_Q(t)$, and a bounded operator $R:\mathbb{R}^k \mapsto S^\perp$ such that for every $x_0 \in \mathcal{H}$ the uncontrolled state trajectory $x(t): = T_A(t)x_0$ and the output $y(t): = Cx(t)$ satisfy

(5.2) $$P_{S^\perp}x(t) = T_Q(t)P_{S^\perp}x_0 + \int_0^t T_Q(t-s)Ry(s)ds, \quad t \geq 0$$

where $P_{S^\perp}$ is the projection on $S^\perp$.

iii)   There exists a generator Q of a C_0–semigroup on $S^\perp$, $T_Q(t)$, and a bounded operator $R:\mathbb{R}^k \mapsto S^\perp$ such that for every $x_0 \in \mathcal{H}$ the uncontrolled frequency domain trajectories $\xi(s) = (s-A)^{-1}x_0$ and $\vartheta(s) = C\xi(s)$; $s \in \rho(A)$ satisfy on an interval $[r,\infty)$

(5.3) $$P_{S^\perp}x_0 = (s-Q)P_{S^\perp}\xi(s) - R\vartheta(s)$$

Proof:

We shall prove successively $i) \Rightarrow ii)$, $ii) \Rightarrow iii)$ and $iii) \Rightarrow i)$.

$i) \Rightarrow ii)$:

Since S is conditioned invariant we have that there exists a bounded operator G such that S is $T_{A+GC}(t)$–invariant. Now define R as $-P_{S^\perp}G$ and define $T_Q(t)$ as $P_{S^\perp}T_{A+GC}(t)P_{S^\perp}$. It is obvious that R and $T_Q(t)$ are bounded operators on $S^\perp$. Furthermore since S is $T_{A+GC}(t)$–invariant we have that for

$z \in S^\perp$, $P_{S^\perp}T_{A+GC}(s)T_{A+GC}(t)z = P_{S^\perp}T_{A+GC}(s)\{P_S T_{A+GC}(t)z + P_{S^\perp}T_{A+GC}(t)z\} = 0 +$

$P_{S^\perp}T_{A+GC}(s)P_{S^\perp}T_{A+GC}(t)z$. Thus $T_Q(t)$ is a semigroup on $S^\perp$. Let Q denote the generator of this semigroup.

From Curtain & Pritchard [9, p.38] we have that

(5.4) $$T_A(t)x_0 = T_{A+GC}(t)x_0 - \int_0^t T_{A+GC}(t-s)GCT_A(s)x_0ds$$

Since S is $T_{A+GC}(t)$–invariant we have that $P_{S^\perp}T_{A+GC}(t)x_0 = P_{S^\perp}T_{A+GC}(t)P_{S^\perp}x_0$ for all x_0 in $\mathcal{H}$. Operation $P_{S^\perp}$ on (5.4) and using this equality gives

(5.5) $$P_{S^\perp}T_A(t)x_0 = P_{S^\perp}T_{A+GC}(t)P_{S^\perp}x_0 - \int_0^t P_{S^\perp}T_{A+GC}(t-s)P_{S^\perp}GCT_A(s)x_0ds$$

With the definitions of $x(.)$, $y(.)$, R and Q this proves the assertion $i) \Rightarrow ii)$.

$ii) \Rightarrow iii)$:

Since the state trajectory $P_{S^\perp}x(t)$ and the "input" trajectory $y(t)$ are exponentially bounded we may take the Laplace transform of equation (5.2). Let $\mathcal{L}$ denote this transform, then

$$\mathcal{L}\left(P_{S^\perp}x(.)\right)(s) = (s-Q)^{-1}P_{S^\perp}x_0 + (s-Q)^{-1}R\,\mathcal{L}\left(y(.)\right)(s) \quad for \; s \in [r,\infty)$$

Since $\mathcal{L}\left(P_{S^\perp}x(.)\right)(s) = \mathcal{L}\left(P_{S^\perp}T_A(.)x_0\right)(s) = P_{S^\perp}(s-A)^{-1}x_0$ and $\mathcal{L}\left(y(.)\right)(s) = C(s-A)^{-1}x_0$ we have proved (5.3).

$iii) \Rightarrow i)$:

We shall prove that $S^\perp$ is frequency invariant for the system (A^*,C^*). Then theorem II.27 and lemma V.3 conclude this assertion.

Let $i_{S^\perp}:S^\perp \mapsto \mathcal{H}$ denote the inclusion mapping i.e. $i_{S^\perp}(z) = z$.

Let $z \in S^\perp$, $x \in \mathcal{H}$ and $s \in [r,\infty) \cap \rho(A) \cap \rho(Q)$. Then $<(s-A^*)^{-1}z,x> = <z,(s-A)^{-1}x> = <z,P_{S^\perp}(s-A)^{-1}x>$ and (5.3) implies that this is equal to

$$<z,i_{S^\perp}(s-Q)^{-1}P_{S^\perp}x> + <z,i_{S^\perp}(s-Q)^{-1}RC(s-A)^{-1}x> =$$
$$<i_{S^\perp}(s-Q^*)^{-1}z,x> + <(s-A^*)^{-1}C^*R^*(s-Q^*)^{-1}z,x>.$$

Since this holds for all $x \in \mathcal{H}$ we have that

$$(s-A^*)^{-1}z = i_{S^\perp}(s-Q^*)^{-1}z + (s-A^*)^{-1}C^*R^*(s-Q^*)^{-1}z; \; s \in [r,\infty) \cap \rho(A) \cap \rho(Q).$$

Thus

(5.6) $$z = (s-A^*)i_{S^\perp}(s-Q^*)^{-1}z + C^*R^*(s-Q^*)^{-1}z,$$

and so z has a strictly proper (ξ,ω) representation with $\xi(s) = i_{S^\perp}(s-Q^*)^{-1}z$ is contained in $S^\perp$. By definition this implies that $S^\perp$ is frequency invariant for the system (A^*,C^*). □

Following Curtain [7] and Schumacher [34] we can combine the concepts of controlled invariance and conditioned invariance to introduce the concept of (C,A,B)-pairs for the system Σ in (5.1).

Definition V.5: (C,A,B)-pair

A pair of closed subspaces (S,V) of the Hilbert space $\mathcal{H}$ with the properties $S \subset V$, S is conditioned invariant and V is controlled invariant is called a (C,A,B) pair.

The concept of (C,A,B)-pair is related to A^e-invariance, where A^e is the extended system operator of the system Σ^e on the extended state space $\mathcal{H}^e = \mathcal{H} \oplus W$, where the Hilbert space W is the state space for the feedback processor. The feedback processor is given by

$$(5.7) \qquad \dot{w} = Nw + My \qquad u = Lw + Ky$$

where M, L, K are bounded linear operators and N is the infinitesimal generator of a strongly continuous semigroup, $T_N(t)$, on W. Defining $x^e = (x,w)$, we obtain the closed loop operator

$$(5.8) \qquad A^e = \begin{bmatrix} A+BKC & BL \\ MC & N \end{bmatrix}$$

A^e is also the infinitesimal generator of a semigroup $T_{A^e}(t)$ since it is a bounded perturbation of diag (A,N) with $D(A^e) = D(A) \oplus D(N)$.

We prove the following modification of the finite dimensional result from Schumacher [34]

Lemma V.6.

Suppose that V^e is a closed $T_{A^e}(t)$-invariant subspace and define the following subspaces

$$(5.9) \qquad S_{orth} := \{x \in \mathcal{H} \mid \begin{bmatrix} x \\ w \end{bmatrix} \in (V^e)^\perp \text{ for some } w \in W\}$$

$$= P_{\mathcal{H}}(V^e)^\perp$$

$$(5.10) \qquad S := (S_{orth})^\perp$$

and

$$V = \{x \in \mathcal{H} : \begin{bmatrix} x \\ w \end{bmatrix} \in V^e \text{ for some } w \in W\}$$

(5.11)

$$:= P_{\mathcal{H}} V^e$$

Then S_{orth} is frequency invariant for the system (A^*, C^*), V is frequency invariant for the system (A, B) and $S = \{x \in \mathcal{H} | \begin{bmatrix} x \\ 0 \end{bmatrix} \in V^e\}$ and is contained in V.

Proof:

We shall begin by showing that V is frequency invariant for the system (A, B). Since V^e is $T_{A^e}(t)$–invariant we have that $(s - A^e)^{-1} V^e \subset V^e$ for all s larger than some $s_0 \in \mathbb{R}$.

Let x be an element of V, then there exists a $w \in W$ such that $\begin{bmatrix} x \\ w \end{bmatrix}$ is an element of V^e.

$$\begin{bmatrix} x \\ w \end{bmatrix} = (s - A^e)(s - A^e)^{-1} \begin{bmatrix} x \\ w \end{bmatrix} = \left\{ s - \begin{bmatrix} A & 0 \\ 0 & N \end{bmatrix} - \begin{bmatrix} BKC & BL \\ MC & 0 \end{bmatrix} \right\} (s - A^e)^{-1} \begin{bmatrix} x \\ w \end{bmatrix} =$$

$$= \left\{ s - \begin{bmatrix} A & 0 \\ 0 & N \end{bmatrix} \right\} (s - A^e)^{-1} \begin{bmatrix} x \\ w \end{bmatrix} - \begin{bmatrix} BKC & BL \\ MC & 0 \end{bmatrix} (s - A^e)^{-1} \begin{bmatrix} x \\ w \end{bmatrix}.$$

So

(5.12) $\quad x = (s - A) P_{\mathcal{H}} (s - A^e)^{-1} \begin{bmatrix} x \\ w \end{bmatrix} - B \left[KC P_{\mathcal{H}} (s - A^e)^{-1} \begin{bmatrix} x \\ w \end{bmatrix} + LP_W (s - A^e)^{-1} \begin{bmatrix} x \\ w \end{bmatrix} \right]$

By the invariance of V^e we have that $(s - A^e)^{-1} \begin{bmatrix} x \\ w \end{bmatrix} \in V^e$, thus $P_{\mathcal{H}} (s - A^e)^{-1} \begin{bmatrix} x \\ w \end{bmatrix}$ is in V.

Furthermore since $\lim_{s \to \infty} s(s - A^e)^{-1} \begin{bmatrix} x \\ w \end{bmatrix} = \begin{bmatrix} x \\ w \end{bmatrix}$ and K, C, $P_{\mathcal{H}}$, L, P_W are bounded operators we have that

(5.13) $\quad \lim_{s \to \infty} s \left\{ KC P_{\mathcal{H}} (s - A^e)^{-1} \begin{bmatrix} x \\ w \end{bmatrix} + LP_W (s - A^e)^{-1} \begin{bmatrix} x \\ w \end{bmatrix} \right\} = KCx + Lw$

So by definition we may conclude that V is frequency invariant for the system (A, B).

Since $V^{e\perp}$ is $T_{A^e}^*(t)$–invariant, we can apply a similar argument to show that S_{orth} is frequency invariant with respect to the system (A^*, C^*). This is left to the reader.

It remains to show that $S = \{x \in \mathcal{H} | \begin{bmatrix} x \\ 0 \end{bmatrix} \in V^e\}$. Let x be an element of S. We shall show that $\begin{bmatrix} x \\ 0 \end{bmatrix} \in \left[V^{e^\perp} \right]^\perp = V^e$. Let $\begin{bmatrix} z \\ w \end{bmatrix}$ be an arbitrary element of $V^{e^\perp}$, then $z \in S_{orth}$, and thus

$$< \begin{bmatrix} x \\ 0 \end{bmatrix}, \begin{bmatrix} z \\ w \end{bmatrix} > = <x,z> = 0$$

Now we shall prove the converse. Let $x \in \mathcal{H}$ be such that $\begin{bmatrix} x \\ 0 \end{bmatrix} \in S$ and $z \in S_{orth}$, then there exists a $w \in W$ such that $\begin{bmatrix} z \\ w \end{bmatrix} \in V^{e^\perp}$. So $0 = < \begin{bmatrix} x \\ 0 \end{bmatrix}, \begin{bmatrix} z \\ w \end{bmatrix} > = <x,z>$, and thus $x \in S_{orth}^\perp$.

$\square$

Remark:

With a similar proof one can show that V is open loop invariant for the system (A,B) and S_{orth} is open loop invariant for the system (A^*,C^*).

Remark:

The projection of a closed linear subspace is not necessarily closed subspace. This is easy to see from the next example.

Let $A: \mathcal{H} \mapsto \mathcal{H}$ be a bounded linear operator with dense range, and this range is unequal to $\mathcal{H}$. Then the graph of A; $V^e := \{(x,Ax)|x \in \mathcal{H}\}$, is closed in $\mathcal{H} \oplus \mathcal{H}$, but the projection of this graph on the second coordinate gives the range of A, which was by assumption not closed.

Remark:

If S_{orth} and V are both closed subspaces, then by theorem II.27 and lemma V.3. (S,V) is a (C,A,B)–pair.

The next lemma will show that if the feedback processor in (5.7) is finite dimensional i.e. W is a finite dimensional space, then S and V defined by (5.10) and (5.11) are closed subspaces and so in this case (S,V) is a (C,A,B)–pair.

Lemma V.7.

Suppose that V^e is a closed $T_{A^e}(t)$–invariant subspace and W is finite dimensional, then the pair (S,V) of subspaces defined by (5.10) and (5.11) is a (C,A,B)–pair and $\dim(V \cap S^\perp) < \infty$.

Proof:

With lemma V.6 and the remark made above, we only have to prove that S_{orth} and V are closed subspaces. We shall begin by showing that V is closed. By definition:

$$V = \{x \in \mathcal{H} | \begin{bmatrix} x \\ w \end{bmatrix} \in V^e, \text{ for some } w \in W\} \text{ and}$$

$$S = \{x \in \mathcal{H} | \begin{bmatrix} x \\ 0 \end{bmatrix} \in V^e\}$$

This equality gives that $S \oplus \{0\} = V^e \cap [\mathcal{H} \oplus \{0\}]$. Since V^e is a closed subspace of $\mathcal{H} \oplus W$ we have that $S \oplus \{0\}$ is a closed subspace, and this easily implies that S is a closed subspace too. So every x in $\mathcal{H}$ can be uniquely decomposed as the sum of two elements with the first in S and the second in $S^{\perp}$. Since S is contained in V, we have that every element v, in V can be uniquely decomposed as

(5.14) $v = s + \tilde{v}$

with $s \in S$ and $\tilde{v} = P_{S^{\perp}}(v) \in V \cap S^{\perp}$. We shall show that the finite dimensionality of W implies the finite dimensionality of $V \cap S^{\perp}$.

Let $\tilde{W}$ be the subspace of W containing all w in W such that there exists an x in $\mathcal{H}$, with $\begin{bmatrix} x \\ w \end{bmatrix} \in V^e$. We shall prove that there exists a linear map from $\tilde{W}$ onto $V \cap S^{\perp}$.

Let $\tilde{w}$ be an element of $\tilde{W}$, then there exists a $x \in \mathcal{H}$ such that $\begin{bmatrix} x \\ \tilde{w} \end{bmatrix} \in V^e$. From the definition of V we get that this x is an element of V and with equation (5.14) we have that $\begin{bmatrix} x \\ \tilde{w} \end{bmatrix} = \begin{bmatrix} s \\ 0 \end{bmatrix} + \begin{bmatrix} \tilde{v} \\ \tilde{w} \end{bmatrix}$; $s \in S; \tilde{v} \in V \cap S^{\perp}$. From the definition of S we have that $\tilde{v}$ is uniquely determined by $\tilde{w}$. Since if $x' \in \mathcal{H}$ is a different ellement with the property that $\begin{bmatrix} x' \\ \tilde{w} \end{bmatrix}$ is an element of V^e, then $\begin{bmatrix} x' \\ \tilde{w} \end{bmatrix} - \begin{bmatrix} x \\ \tilde{w} \end{bmatrix} = \begin{bmatrix} x'-x \\ 0 \end{bmatrix} \in V^e$, thus $x' - x \in S$ or $P_{S^{\perp}}x' = P_{S^{\perp}}x = \tilde{v}$.

So there exists a linear map from $\tilde{W}$ to $V \cap S^{\perp}$, and by the definition of V this map is onto. Thus

(5.15) $dim(V \cap S^{\perp}) \leq dim(\tilde{W}) \leq dim(W) < \infty$

or $V \cap S^{\perp}$ is finite dimensional. Since $V = S \oplus (V \cap S^{\perp})$ we have that V is a closed subspace.

A similar argument as that used above can be used to show that $S_{orth} = \{x \in \mathcal{H} | \exists w \in W, \begin{bmatrix} x \\ w \end{bmatrix} \in V^{e^{\perp}}\}$ is a closed subspace too.   □

This result shows that a semigroup invariant subspace for the extended system introduces a (C,A,B)–pair. We shall now show that conversely every (C,A,B)–pair brings on an extended system and its associated invariant subspace.

Before we can prove this result we need the following result of Curtain [7].

Lemma V.8.

Given a (C,A,B)–pair, (S,V), there exist $F \in \mathcal{L}(\mathcal{H},\mathbb{R}^m)$, $G \in \mathcal{L}(\mathbb{R}^k,\mathcal{H})$, and $K \in \mathcal{L}(\mathbb{R}^k,\mathbb{R}^m)$ such that

i) $T_{A+BF}(t)V \subset V$, $T_{A+GC}(t)S \subset S$,

ii) $F = KC + F_0$, $G = BK + G_0$ and

iii) $Ker \, F_0 \supset S$ and $Im \, G_0 \subset V$

Proof:

See Curtain [7]. $\square$

Lemma V.9.

If (S,V) is a (C,A,B)–pair, then there exists an extension space $W \cong V/S$, bounded linear operators M, L, K and a generator N of a C_0–semigroup $T_N(t)$ on W and a closed subspace V^e of $\mathcal{H}^e := \mathcal{H} \oplus W$ which is invariant under the semigroup generated by A^e. Furthermore the subspace V^e satisfies $V = P_{\mathcal{H}} V^e$ and $S = V^e \cap \{\mathcal{H} \oplus \{0\}\}$.

Proof:

Since V and S are closed and $V \supset S$, we have the representation $V \cong S \oplus V/S$. Let $W \cong V/S$ and define $\mathcal{H}^e = \mathcal{H} \oplus W$. Let R be a bounded linear map from V to W such that $Ker \, R = S$ and $Im \, R = W$. Then there exists a map $R^+ \in \mathcal{L}(W,V)$ such that $R R^+ = I_W$. Furthermore this mapping satisfies $R^+ Rx = 0$ if and only if $x \in S$.

Define the subspace V^e of $\mathcal{H}^e$ by

(5.16) $V^e = \left\{ \begin{pmatrix} x \\ Rx \end{pmatrix} \mid x \in V \right\}$

By definition $S = V^e \cap \mathcal{H}$ and $V = P_{\mathcal{H}} V^e$. Let F, G, F_0 and G_0 be the same as in lemma V.8. We shall investigate the semigroup $T_{A+BF+G_0C}(t)$ more closely.

a) $T_{A+BF+G_0C}(t)V \subset V$

This follows easily from Lemma II.2. with lemma V.8. i), iii).

b) $\quad T_{A+BF+G_0C}(t)s = T_{A+GC}(t)s \in S$; for all s in S.

This is clear from lemma V.1. and lemma V.8.

Now we shall consider the operator

$$T_N(t) := R \ T_{A+BF+G_0C}(t)R^+.$$

By a) this is a well defined operator on W. Furthermore since R^+Rx-x is an element of S, for $x \in V$, we have by b) that

(5.17) $\quad R\left[T_{A+BF+G_0C}(t)R^+Rx - T_{A+BF+G_0C}(t)x\right] \subset R(S) = \{0\}$, for all x in V.

So $\quad T_N(t)Rx := T_{A+BF+G_0C}(t)R^+Rx = T_{A+BF+G_0C}(t)x$, for all x in V, and

$$T_N(t)T_N(s) = T_N(t)R \ T_{A+BF+G_0C}(s)R^+ = R \ T_{A+BF+G_0C}(t)T_{A+BF+G_0C}(s)R^+ =$$

$$R \ T_{A+BF+G_0C}(t+s)R^+ = T_N(t+s),$$

and thus $T_N(.)$ is C_0–semigroup on W.

Following Curtain [7] and Schumacher [34] we construct N, M and L as follows

(5.18) $\quad \begin{cases} M=-R \ G_0, \ L=F_0R^+ \\ N \text{ is the generator of } T_N(t) \end{cases}$

Clearly $M \in \mathcal{L}(\mathbf{R}^k, W)$, $L \in \mathcal{L}(W, \mathbf{R}^m)$.

We shall show that V^e is $T_{A^e}(t)$–invariant ($T_{A^e}(t)$ is the C_0–semigroup generated by A^e). Let $\begin{bmatrix} x \\ Rx \end{bmatrix}$ be an element of V^e, then

$$\begin{bmatrix} T_{A+BF+G_0C}(t) & 0 \\ 0 & T_N(t) \end{bmatrix} \begin{bmatrix} x \\ Rx \end{bmatrix} = \begin{bmatrix} T_{A+BF+G_0C}(t)x \\ RT_{A+BF+G_0C}(t)R^+Rx \end{bmatrix} \qquad (5.12)$$

$$= \begin{bmatrix} T_{A+BF+G_0C}(t)x \\ RT_{A+BF+G_0C}(t)x \end{bmatrix} \in V^e \text{ by a).}$$

Thus V^e is invariant under the semigroup generated by

$$\begin{pmatrix} A+BF+G_0C & 0 \\ 0 & N \end{pmatrix} = \begin{pmatrix} A+BKC+BF_0+G_0C & 0 \\ 0 & N \end{pmatrix} =$$

$$\begin{pmatrix} A+BKC & BL \\ MC & N \end{pmatrix} + \begin{pmatrix} BF_0+G_0C & -BL \\ -MC & 0 \end{pmatrix} = A^e + \begin{pmatrix} BF_0+G_0C & -BL \\ -MC & 0 \end{pmatrix}.$$

From lemma II.2. we have that V^e is $T_{A^e}(t)$ invariant if and only if

$$\begin{pmatrix} BF_0+G_0C & -BL \\ -MC & 0 \end{pmatrix} \begin{pmatrix} x \\ Rx \end{pmatrix} \in V^e \text{ for all } x \text{ in } V. \text{ This assertion will be proved using}$$

the other properties of G_0 and F_0.

$$\begin{pmatrix} BF_0+G_0C & -BL \\ -MC & 0 \end{pmatrix} \begin{pmatrix} x \\ Rx \end{pmatrix} = \begin{pmatrix} B(F_0x - LRx)+G_0Cx \\ -MCx \end{pmatrix} =$$

$$= \begin{pmatrix} B(F_0x - F_0R^+Rx)+G_0Cx \\ + RG_0Cx \end{pmatrix}. \text{ Since } S \subset Ker\, F_0 \text{ we have that this equals}$$

$$\begin{pmatrix} G_0Cx \\ RG_0Cx \end{pmatrix}, \text{ and by the inclusion } Im\, G_0 \subset V, \text{ this is an element of } V^e. \qquad \square$$

Section V.2: Disturbance Decoupling Problem with Measurement Feedback

We now consider the disturbance decoupling problem with measurement feedback, DDPM which was the subject of Schumacher [34], Willems & Commault [41] and Curtain [7]. Consider the system on the Hilbert space $\mathcal{H}$.

$$\dot{x} = Ax + Bu + Eq$$

(5.19)

$$y = Cx, \quad z = Dx$$

where A, B, C are as in section V.1. and $E \in \mathcal{L}(R^q, \mathcal{H})$, $D \in \mathcal{L}(\mathcal{H}, R^d)$, q is the disturbance input and z the output to be decoupled.

DDPM is to design a feedback processor of the form (5.7) such that the output z is decoupled from the disturbance q, i.e.

$$\text{(5.20)} \qquad D^e \int_0^t T_{A^e}(t-s)E^e q(s)ds = 0, \quad t \geq 0$$

where $T_{A^e}(t)$ is the semigroup generated by the closed–loop system operator A^e, (5.8) $D^e = (D,0)$ and $E^e = \begin{bmatrix} E \\ 0 \end{bmatrix}$. Pictorially our configuration is given in figure 5.1.

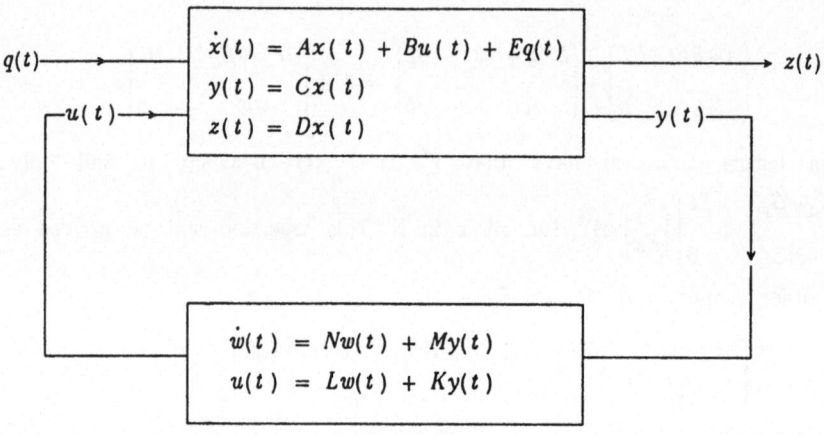

q(t) →

$$\dot{x}(t) = Ax(t) + Bu(t) + Eq(t)$$
$$y(t) = Cx(t)$$
$$z(t) = Dx(t)$$

→ z(t)

—u(t)→

—y(t)—

$$\dot{w}(t) = Nw(t) + My(t)$$
$$u(t) = Lw(t) + Ky(t)$$

figure 5.1

Before we shall prove the DDPM in terms of subspaces we shall prove it in terms of transfer functions.

Theorem V.10.

The DDPM is solvable with the compensator (N,M,L,K) if and only if there exists a $X(s) \in \big(\mathcal{L}(\mathcal{Y},\mathcal{U})\big)(s)$ such that $\lim_{s \to \infty} X(s)$ exists,

(5.21) $D(s-A)^{-1}E + D(s-A)^{-1}B X(s) C(s-A)^{-1}E \equiv 0$ on $[r,\infty)$

and $X(s)\left[\ I + C(s-A)^{-1}B X(s)\ \right]^{-1} = K + L(s-N)^{-1}M$

Remark:

Since $C(s-A)^{-1}B$ is strictly proper and $X(.)$ is proper we have that on an interval $[r_1,\infty)$ the inverse in the last line of this theorem always exists.

Proof:

We shall begin with deriving fundamental equalities.

In order to derive equation (5.21) we shall need the following equalities on $\rho(A) \cap \rho(A^e)$

$$(s-A^e)^{-1} = \begin{bmatrix} (s-A)^{-1} & 0 \\ 0 & (s-N)^{-1} \end{bmatrix} + (s-A^e)^{-1} \begin{bmatrix} BKC \ BL \\ MC \ 0 \end{bmatrix} \begin{bmatrix} (s-A)^{-1} & 0 \\ 0 & (s-N)^{-1} \end{bmatrix}$$

(5.22)

$$(s-A^e)^{-1} = \begin{bmatrix} (s-A)^{-1} & 0 \\ 0 & (s-N)^{-1} \end{bmatrix} + \begin{bmatrix} (s-A)^{-1} & 0 \\ 0 & (s-N)^{-1} \end{bmatrix} \begin{bmatrix} BKC & BL \\ MC & 0 \end{bmatrix} (s-A^e)^{-1}$$

So

$$D^e(s-A^e)^{-1}E^e = D^e \left\{ \begin{bmatrix} (s-A)^{-1} & 0 \\ 0 & (s-N)^{-1} \end{bmatrix} + \begin{bmatrix} (s-A)^{-1} & 0 \\ 0 & (s-N)^{-1} \end{bmatrix} \begin{bmatrix} BKC & BL \\ MC & 0 \end{bmatrix} (s-A^e)^{-1} \right\} E^e$$

$$= D(s-A)^{-1}E + \begin{pmatrix} D(s-A)^{-1}, & 0 \end{pmatrix} \begin{bmatrix} BKC & BL \\ MC & 0 \end{bmatrix} (s-A^e)^{-1}E^e =$$

$$= D(s-A)^{-1}E + D(s-A)^{-1}B \begin{pmatrix} KC, & L \end{pmatrix} (s-A^e)^{-1}E^e = D(s-A)^{-1}E +$$

$$D(s-A)^{-1}B \begin{pmatrix} KC, & L \end{pmatrix} \left\{ \begin{bmatrix} (s-A)^{-1} & 0 \\ 0 & (s-N)^{-1} \end{bmatrix} + (s-A^e)^{-1} \begin{bmatrix} BKC \ BL \\ MC \ 0 \end{bmatrix} \begin{bmatrix} (s-A)^{-1} & 0 \\ 0 & (s-N)^{-1} \end{bmatrix} \right\} E^e$$

$$= D(s-A)^{-1}E + D(s-A)^{-1}B \left\{ K + \begin{pmatrix} KC, & L \end{pmatrix} (s-A^e)^{-1} \begin{bmatrix} BK \\ M \end{bmatrix} \right\} C(s-A)^{-1}E.$$

So if $X(s)$ denotes $\left\{ K + \begin{pmatrix} KC, & L \end{pmatrix} (s-A^e)^{-1} \begin{bmatrix} BK \\ M \end{bmatrix} \right\}$, then we have that

(5.23) $D^e(s-A^e)^{-1}E^e = D(s-A)^{-1}E + D(s-A)^{-1}B X(s) C(s-A)^{-1}E.$

Furthermore we have that $X(s)$ is with (5.22) equal to

$$K + \begin{pmatrix} KC,L \end{pmatrix} \left\{ \begin{bmatrix} (s-A)^{-1} & 0 \\ 0 & (s-N)^{-1} \end{bmatrix} + \begin{bmatrix} (s-A)^{-1} & 0 \\ 0 & (s-N)^{-1} \end{bmatrix} \begin{bmatrix} BKC \ BL \\ MC \ 0 \end{bmatrix} (s-A^e)^{-1} \right\} \begin{bmatrix} BK \\ M \end{bmatrix} =$$

$$= K + KC(s-A)^{-1}BK + L(s-N)^{-1}M + \begin{pmatrix} KC(s-A)^{-1}, & L(s-N)^{-1} \end{pmatrix} \begin{bmatrix} BKC \ BL \\ MC \ 0 \end{bmatrix} (s-A^e)^{-1} \begin{bmatrix} BK \\ M \end{bmatrix} =$$

$$= K + L(s-N)^{-1}M + KC(s-A)^{-1}BK + KC(s-A)^{-1}B\,(KC,L)\,(s-A^e)^{-1}\begin{pmatrix}BK\\M\end{pmatrix} +$$

$$+ L(s-N)^{-1}M\,(C,0)\,(s-A^e)^{-1}\begin{pmatrix}BK\\M\end{pmatrix} =$$

$$= F(s) + KC(s-A)^{-1}B\,X(s) + L(s-N)^{-1}M\,(C,0)\,(s-A^e)^{-1}\begin{pmatrix}BK\\M\end{pmatrix}$$

where $F(s)$ is the transfer function of the compensator so $F(s) = K + L(s-N)^{-1}M$. Using a similar argument as above one can show that

$$(C,0)\,(s-A^e)^{-1}\begin{pmatrix}BK\\M\end{pmatrix} = C(s-A)^{-1}B\,X(s)$$

So $X(s) = F(s) + KC(s-A)^{-1}B\,X(s) + L(s-N)^{-1}M\,(C,0)\,(s-A^e)^{-1}\begin{pmatrix}BK\\M\end{pmatrix} = F(s) +$

$$KC(s-A)^{-1}B\,X(s) + L(s-N)^{-1}M\,C(s-A)^{-1}B\,X(s) = F(s)\left[\,I + C(s-A)^{-1}B\,X(s)\,\right].$$

Since $\lim_{s\to\infty} X(s) = K$ and $\lim_{s\to\infty} C(s-A)^{-1}B = 0$ we have that on an interval $[r_0,\infty)$ the inverse of $\left[I + C(s-A)^{-1}B\,X(s)\right]$ exists, and so

$$(5.24) \qquad X(s)\left[I + C(s-A)^{-1}B\,X(s)\,\right]^{-1} = F(s) = K + L(s-N)^{-1}M.$$

The equalities (5.23) and (5.24) prove the theorem.

$\square$

If we do not know that the solution $X(.)$ of (5.21) is related to the realisation (N,M,L,K) by (5.24), then we must assume conditions on the system (5.1). Furthermore as for the DDP in chapter 3 to obtain necessary and sufficient conditions for DDPM in terms of invariant subspaces, we need to assume that the largest frequency invariant subspace with respect to (A,B) in $Ker\,D$, $V_{(A,B)}(Ker\,D)$ is closed and furthermore that the same holds for $V_{(A^*,C^*)}(Ker\,E^*)$. Then we have from chapter 3 that the largest controlled invariant subspace contained in $Ker\,D$, $V^*(Ker\,D)$, exists as well as the smallest conditioned invariant subspace containing $Im\,E$, $S^*(Im\,E)$, and $V_{(A,B)}(Ker\,D) = V^*(Ker\,D)$, $\{V_{(A^*,C^*)}(Ker\,E^*)\}^{\perp} = S^*(Im\,E)$.

Theorem V.11.

If $V_{(A,B)}(Ker\ D)$ and $V_{(A^*,C^*)}(Ker\ E^*)$ are closed subspaces, then the following assertions are equivalent

i)   DDPM is solvable

ii) $S^*(Im\ E) \subset V^*(Ker\ D)$

where $S^*(Im\ E) = (V_{(A^*,C^*)}(Ker\ E^*))^\perp$ and $V^*(Ker\ D) = V_{(A,B)}(Ker\ D)$.

iii)   There exists a $X(s) \in (\mathcal{L}(\mathcal{Y},\mathcal{U}))(s)$ such that $\lim_{s \to \infty} X(s)$ exists and

$$D(s-A)^{-1}E + D(s-A)^{-1}B\ X(s)\ C(s-A)^{-1}E \equiv 0 \ \ on \ [r,\infty)$$

Proof:

We shall prove $i)\leftrightarrow ii)$, $i)\Rightarrow iii)$ and $iii)\Rightarrow ii)$.

Furthermore we shall use S^* and V^* instead of $S^*(Im\ E)$ and $V^*(Ker\ D)$, respectively.

$i)\Rightarrow ii)$:

If DDPM is solvable, then there exists a subspace V^e that is $T_{A^e}(t)$–invariant, $Im\ E^e \subset V^e \subset Ker\ D^e$, namely the space containing all x^e such that $x^e = \int_0^t T^e(t-s)E^e q(s)ds$ for some t and $q(.)$.

From lemma V.6. we have that

$$S_{orth} := \{x \in \mathcal{H} \mid \begin{bmatrix} x \\ w \end{bmatrix} \in V^{e\perp} \text{ for some } w \in W\}$$

is frequency invariant for the system (A^*,C^*) and by the definition of V^e, $S_{orth} \subset Ker\ E^*$. So $S_{orth} \subset V_{(A^*,C^*)}(Ker\ E^*)$, and by definition of S and S^* we have

(5.25) $S^* = [V_{(A^*,C^*)}(Ker\ E^*)]^\perp \subset S_{orth}^\perp = S.$

Furthermore we have that

$$V := \{x \in \mathcal{H} \mid \begin{bmatrix} x \\ w \end{bmatrix} \in V^e \text{ for some } w \in W\}$$

is frequency invariant, and by definition we have that $V \subset Ker\ D$.

So we have that

(5.26) $S^* \subset S \subset V \subset V^*$

ii)⇒i)

Since $V_{(A,B)}(Ker\,D)$ and $V_{(A^*,C^*)}(Ker\,E^*)$ are closed, we have with **ii)** that (S^*, V^*) is a (C,A,B)–pair.

By lemma V.9. we can construct an extended system operator A^e on $\mathcal{H}^e = \mathcal{H} \oplus W$, $W \cong V^*/S^*$ and a subspace V^e which is $T_{A^e}(t)$–invariant. Consider the extended system Σ^e on $\mathcal{H}^e$

$$\dot{x}^e = A^e x^e + B^e u^e + E^e q$$

(5.27)

$$y = C^e x^e, \quad z = D^e x^e$$

where $x^e = \begin{bmatrix} x \\ w \end{bmatrix}$, $B^e = \begin{bmatrix} B \\ 0 \end{bmatrix}$, $C^e = (C, 0)$.

Now $S^* \supset Im\,E$ and $V^* \subset Ker\,D$ implies that

(5.28) $Im\,E^e \subset V^e \subset Ker\,D^e$

Now V^e is $T_{A^e}(t)$–invariant and so from (5.28) we obtain

$$T_{A^e}(t)Im\,E^e \subset T_{A^e}(t)V^e \subset V^e \subset Ker\,D^e$$

and (5.20) holds.

i)⇒iii):

This is a direct consequence of theorem V.10.

iii)⇒ii):

Since we do not know that $X(s)\left[\ I + C(s-A)^{-1}B\,X(s)\ \right]^{-1}$ has a realisation (N, M, L, K) we cannot apply theorem V.10.

Let $X(s)$ be the solution of (5.21), then we shall define two subspaces, V and S_{orth}, of $\mathcal{H}$ such that $Im\,E \subset S_{orth}^{\perp} \subset V \subset Ker\,D$, V is frequency invariant for the system (A,B) and S_{orth} is frequency invariant for the system (A^*, C^*).

We note that equation (5.21) always implies that $Im\,E \subset Ker\,D$.

Now define V as

(5.29) $V = \{x \in \mathcal{H} |$ there exists $s \in [r,\infty)$ and $q \in Q$ such that

$$x = (s-A)^{-1}Eq + (s-A)^{-1}B\,X(s)\,C(s-A)^{-1}Eq\}$$

Now we trivially have that V is contained in $Ker\,D$. Thus it remains to show the frequency invariance.

Let s be an arbitrary element of $[r,\infty)$ and let x be an element of V, then

(5.30) $(s-A)^{-1}x = (s-A)^{-1}\Big\{(s_0-A)^{-1}Eq + (s_0-A)^{-1}B\,X(s_0)\,C(s_0-A)^{-1}Eq\Big\}$;   for

some $s_0 \in [r,\infty)$ and $q \in Q$.

Using the resolvent equation, equation (5.30) gives:

(5.31) $(s-A)^{-1}x = \dfrac{1}{(s_0-s)}\Big\{(s_0-A)^{-1}Eq + (s_0-A)^{-1}B\,X(s_0)\,C(s_0-A)^{-1}Eq\Big\} +$

$\qquad\qquad \dfrac{1}{(s_0-s)}\Big\{(s-A)^{-1}Eq + (s-A)^{-1}B\,X(s_0)\,C(s_0-A)^{-1}Eq\Big\} =$

$\qquad = \dfrac{1}{(s_0-s)}x + \dfrac{1}{(s_0-s)}(s-A)^{-1}Eq + (s-A)^{-1}B\,\dfrac{1}{(s_0-s)}\Big\{X(s_0)\,C(s_0-A)^{-1}Eq\Big\}$

$\qquad = \dfrac{1}{(s_0-s)}x + \dfrac{1}{(s_0-s)}\Big\{(s-A)^{-1}Eq + (s-A)^{-1}B\,X(s)\,C(s-A)^{-1}Eq\Big\} +$

$\qquad + (s-A)^{-1}B\,\dfrac{1}{(s_0-s)}\Big\{X(s_0)\,C(s_0-A)^{-1}Eq - X(s)\,C(s-A)^{-1}Eq\Big\}$

Defining $\xi(s)$ as

(5.32) $\xi(s) = \dfrac{1}{(s_0-s)}x + \dfrac{1}{(s_0-s)}\Big\{(s-A)^{-1}Eq + (s-A)^{-1}B\,X(s)\,C(s-A)^{-1}Eq\Big\}$

and $\omega(s)$ as

(5.33) $\omega(s) = \dfrac{1}{(s_0-s)}\Big\{X(s_0)\,C(s_0-A)^{-1}Eq - X(s)\,C(s-A)^{-1}Eq\Big\}$

Then $\xi(s)$ is contained in V, $\omega(s)$ is strictly proper and equation (5.31) gives that this $\xi(.)$ and $\omega(.)$ is a (ξ,ω)–representation of x. Thus V is frequency invariant for the system (A,B).

    For S_{orth} we take the largest frequency invariant subspace with respect to (A^*,C^*), contained in $Ker\,E^*$. Then we only have to check that $V^\perp \subset S_{orth}$. Define for $x \in V^\perp$ and $s \in [r,\infty)$

$$\omega(s) = X^*(s)B^*(s-A^*)^{-1}x$$

(5.34)

$$\xi(s) = (s-A^*)^{-1}x + (s-A^*)^{-1}C^*\omega(s)$$

Then this is a strictly proper (ξ,ω)–representation of x, with respect to the system (A^*,C^*). In order to prove that x is an element of S_{orth} it is sufficient to prove that $\xi(.) \in Ker\, E^*$, see definition III.4.

Let q be an arbitrary element of Q, then

$$<E^*\xi(s),q> = \; <(s-A^*)^{-1}x + (s-A^*)^{-1}C^*\omega(s), Eq> \; =$$

$$<(s-A^*)^{-1}x + (s-A^*)^{-1}C^*X^*(s)B^*(s-A^*)^{-1}x, Eq> \; =$$

$$<x,(s-A)^{-1}Eq \; + \; (s-A)^{-1}B\,X(s)\,C(s-A)^{-1}Eq> \; = 0, \; since \; x \in V^\perp$$

Since q was an arbitrary element of Q, we have that $E^*\xi(s) = 0$.

So summarizing we have the existence of a pair of subspace, S_{orth} and V, such that

V is frequency invariant for the system (A,B) and $V \subset Ker\, D$,

S_{orth} is frequency invariant for the system (A^*,C^*) and $S_{orth} \subset Ker\, E^*$,

$S_{orth}^\perp \subset V$

Combining these results give that

$$S^*(Im\, E) = S_{orth}^\perp \subset V \subset V^*(Ker\, D)$$

□

Now we shall investigate under which conditions we obtain the existence of a finite dimensional compensator.

Theorem V.12.

The following assertions are equivalent

i)   DDPM is solvable with a finite dimensional compensator

ii)   There exists a (C,A,B)–pair (S,V) with $dim(V \cap S^\perp) < \infty$ and

$$Im\, E \subset S \subset V \subset Ker\, D$$

iii)   There exists a $X(s)\in\left(\mathcal{L}(\mathcal{Y},\mathcal{U})\right)(s)$ such that $\lim_{s\to\infty} X(s)$ exists,

(5.35) $$D(s-A)^{-1}E + D(s-A)^{-1}B\,X(s)\,C(s-A)^{-1}E \equiv 0 \ on\ [\tau,\infty)$$

and $F(s):=X(s)\left[\,I + C(s-A)^{-1}B\,X(s)\,\right]^{-1}$ is a proper rational function.

Proof:

i)$\Rightarrow$ii):

Let V^e be the closure of the space consisting of all x^e such that
$x^e=\int_0^t T_{A^e}(t-s)E^e q(s)ds$ for some t and $q(s)$.
This subspace is $T_{A^e}(t)$–invariant and contained in the kernel of D^e. By lemma V.7. the pair (S,V) is a (C,A,B)–pair, where

$$S=\{x\in\mathcal{H}|\begin{pmatrix}x\\0\end{pmatrix}\in V^e\}$$

$$V=\{x\in\mathcal{H}|\begin{pmatrix}x\\w\end{pmatrix}\in V^e,\ for\ some\ w\in W\}$$

and furthermore $dim(V\cap S^\perp)<\infty$.
From the definition of S and V we have

$$Im\ E\subset S\subset V\subset Ker\ D.$$

ii)$\Rightarrow$i)

From lemma V.9 we have the existence of a finite dimensional compensator and a subspace $V^e\subset\mathcal{H}\oplus W$ which is invariant under the extended system $T_{A^e}(t)$.
Furthermore we have that $S\oplus\{0\}\subset V^e\subset V\oplus W$. So $Im\ E\subset S$ and $V\subset Ker\ D$ implies that

$$Im\ E^e\subset V^e\subset Ker\ D^e$$

Thus $T_{A^e}(t)Im\ E^e\subset T_{A^e}(t)V^e\subset V^e\subset Ker\ D^e$, and so i) holds.

i)$\Leftrightarrow$iii)

This is a direct consequence of theorem V.10.   □

Remark:

As in finite dimensions we have that the minimal compensator state dimension is:

$$min\{dim(V/S)\,|\,(S,V) \text{ is a } (C,A,B)\text{–pair} \quad \text{with} \quad Im\,E \subset S \subset V \subset Ker\,D\}.$$

So direct output feedback is possible if and only if there exists a V which is controlled **and** conditioned invariant with $Im\,E \subset V \subset Ker\,D$.

CHAPTER VI: THE DISTURBANCE DECOUPLING PROBLEM WITH MEASUREMENT FEEDBACK AND STABILITY

As in the previous chapter we shall derive necessary and sufficient conditions such that the Disturbance Decoupling Problem with Measurement feedback and Stability is solvable and once again the main aim is to obtain a finite dimensional compensator. Whereas in the previous chapter we closely followed the work of Curtain [7] and thus of Schumacher [34], now we shall follow the line in Willems and Commault [41].

Again we shall study the system (5.1) on a Hilbert space $\mathcal{H}$, which was given by

$$(6.1) \qquad \Sigma: \quad \begin{aligned} \dot{x} &= Ax + Bu \\ y &= Cx \end{aligned}$$

where A is the infinitesimal generator of a strongly continuous semigroup $T_A(t)$ on $\mathcal{H}$, $B \in \mathcal{L}(\mathbb{R}^m, \mathcal{H})$ and $C \in \mathcal{L}(\mathcal{H}, \mathbb{R}^k)$.

We shall start with refering briefly the stability theory for infinite dimensional systems.

Section VI.1. Stability, Stabilizability and Stabilizability Subspaces.

In this section we shall introduce the notion of stability and to this concept we shall associate a new subspace. With the basic results derived in this section we can give a geometric characterization of the solvability of DDPMS.

We remark that in this monograph we shall consider only exponential stability.

Definition VI.1: Stability.
We say that $T_A(t)$ or A is stable if and only if there exists a $\delta > 0$ and $M \geq 1$ such that

$$(6.2) \qquad \|T_A(t)\| \leq M e^{-\delta t}$$

Definition VI.2: Stabilizability

The system (A,B) is stabilizable if there exists a bounded feedback law F such that $A+BF$ is stable.

From Jacobson & Nett [21], Nefedov & Sholokhovich [25] and Datko [13] we have the following characterization of stabilizable systems.

Theorem VI.3.

For the system (6.1) the following assertions are equivalent

i) The system (A,B) is stabilizable

ii) There exists a $\delta > 0$ such that the state space $\mathcal{H}$ can be decomposed as $\mathcal{H} = \mathcal{H}_u \oplus \mathcal{H}_s$ which are both $T_A(t)$–invariant, $\mathcal{H}_u$ is finite dimensional,

$$\sigma(A|_{\mathcal{H}_u}) \subset \{s \in \mathbb{C} \mid Re\ s > -\delta\}, \quad \|T_A(t)|_{\mathcal{H}_s}\| \le M e^{-\delta t}$$

and the system restricted to $\mathcal{H}_u$ is controllable.

iii) For all x_0 in $\mathcal{H}$, there exists a $u(.) \in L^2([0,\infty);\mathcal{U})$ such that

$$x(t) := T_A(t)x_0 + \int_0^t T_A(t-s)Bu(s)ds$$

satisfies $x(.) \in L^2([0,\infty);\mathcal{H})$.

iv) For all x_0 in $\mathcal{H}$ there exists a $\omega(.) \in H^2(\mathcal{U})$ and a $\xi(.) \in H^2(D(A))$ such that

$$x_0 = (s-A)\xi(s)\ -\ B\omega(s)$$

holds for all $s \in \mathbb{C}_+ := \{s \in \mathbb{C} \mid Re\ s > 0\}$.

Proof:

$i) \leftrightarrow ii)$, see Jacobson & Nett [21] or Nefedov & Sholokhovich [25].

$i) \leftrightarrow iii)$, see Datko [13].

$iii) \leftrightarrow iv)$, This is a direct consequence of the fact that $H^2(\mathcal{U})$ is isomorphic to $L^2([0,\infty);\mathcal{U})$. □

From this result we have the following corollary.

Corollary VI.4.

Suppose that the system (A,B) is stabilizable. Then we have the following results.

i) $T_A(t)$ is stable if and only if there exists a $\delta > 0$ such that $\sigma(A) \subset \{s \in \mathbb{C} | Re\ s < -\delta\}$.

ii) Let $B' \in \mathcal{L}(\mathbb{R}^p; \mathcal{H})$. Then the system (A,B') is stabilizable if and only if the system $(A|_{\tilde{\mathcal{H}}_u}, P_{\tilde{\mathcal{H}}_u} B')$ is controllable, where $\tilde{\mathcal{H}}_u$ is the span of all (generalized) unstable eigenvectors.

Remark:

i) of Corollary VI.4. does not necessarily imply that A satisfies the spectrum determined growth assumption.

Proof:

i)

if): Since the system (A,B) is stabilizable, there exist $T_A(t)$-invariant subspaces $\mathcal{H}_s$ and $\mathcal{H}_u$ such that $\dim(\mathcal{H}_u) < \infty$ and $\sigma(A|_{\mathcal{H}_u}) \subset \mathbb{C}_{-\delta,+} := \{s \in \mathbb{C} | Re\ s > -\delta\}$ for some $\delta > 0$. Since $\mathcal{H}_u$ is a finite dimensional $T_A(t)$-invariant subspace we have from lemma I.9 that $\sigma(A|_{\mathcal{H}_u}) \subset \sigma(A) \subset \mathbb{C}_{-\delta,-} := \{s \in \mathbb{C} | Re\ s < -\delta\}$. So $\mathcal{H}_u = \{0\}$ and thus $\mathcal{H} = \mathcal{H}_s$, and VI.3.ii) gives the desired result.

only if): See Curtain and Pritchard [9, p.17].

ii)

Since (A,B) is stabilizable we have that the dimension of the space of all generalized unstable eigenvectors is finite dimensional. So there exists a $\tilde{\mathcal{H}}_s$ which is $T_A(t)$-invariant such that $\mathcal{H} = \tilde{\mathcal{H}}_u \oplus \tilde{\mathcal{H}}_s$ and $\|T_A(t)|_{\tilde{\mathcal{H}}_s}\| \leq Me^{-\delta t}$ for some $M \geq 1$ and $\delta > 0$. $\tilde{\mathcal{H}}_s$ is $\mathcal{H}_s$ plus all stable (generalized) eigenvectors in $\mathcal{H}_u$. The result of *ii)* follows now easily from theorem VI.6.

The next result will show that a similar decomposition as in *ii)* of theorem VI.3 holds for all semigroup invariant subspaces.

Lemma VI.5.

Suppose that (A,B) is stabilizable. Then if V is a closed $T_A(t)$-invariant subspace, then V can be decomposed in

$$V = V_u \oplus V_s$$

where V_u and V_s are both $T_A(t)$-invariant, $\dim(V_u) < \infty$ and $T_A(t)|_{V_s}$ is stable.

Proof:

Since (A,B) is stabilizable we can decompose the state space $\mathcal{H}$ in $\mathcal{H}_u$ and $\mathcal{H}_s$ such that $\mathcal{H}_u$ and $\mathcal{H}_s$ are both $T_A(t)$-invariant, $\mathcal{H}_u$ is finite dimensional and $T_A(t)|_{\mathcal{H}_s}$ is stable. Let $P_u = \frac{1}{2\pi i}\int_\Gamma (\lambda - A)^{-1}d\lambda$, where Γ is a jordan curve which encircles only the eigenvalues of $A|_{\mathcal{H}_u}$. Then P_u and $P_s := I - P_u$ are respectively the projection on $\mathcal{H}_u$ along $\mathcal{H}_s$ and the projection on $\mathcal{H}_s$ along $\mathcal{H}_u$. Since $\mathcal{H}_u$ and $\mathcal{H}_s$ are both $T_A(t)$-invariant we have that

$$(6.3) \qquad P_u T_A(t) = T_A(t) P_u.$$

Define $V_u := P_u V$ and $V_s := P_s V$, then $\dim(V_u) < \infty$. Furthermore the invariance of V together with (6.3) implies that V_u and V_s are both $T_A(t)$-invariant. With a similar argument as in the only if part of theorem IV.6 we can prove that $V_u \subset V$, and this implies that $V = V_u \oplus V_s$. It remains to prove the stability of $T_A(t)|_{V_s}$, but this follows trivially from the stability of $T_A(t)|_{\mathcal{H}_s}$. $\qquad\square$

We define detectability as the dual concept of stabilizability.

Definition VI.6: Detectable.

This system (C,A) is detectable if there exists an operator $G \in \mathcal{L}(\mathbb{R}^k, \mathcal{H})$ such that $A + GC$ is stable.

Definition VI.7: Stabilizable-Detectable.

The system (6.1) is stabilizable-detectable if (A,B) is stabilizable and (C,A) is detectable.

As for finite dimensional systems we have that joint stabilizability and detectability is necessary and sufficient for the existence of a finite dimensional compensator such that the closed loop system A^e is stable. A^e is the extended system operator of the system Σ^e on the extended state space $\mathcal{H}^e = \mathcal{H} \oplus W$, where the finite dimensional space W is the state space for the feedback processor. The feedback processor is given by

$$(6.4) \qquad \dot{w} = Nw + My \qquad u = Lw + Ky$$

where N, M, L, K are matrices. Defining $x^e = (x,w)$, we obtain the closed loop operator

(6.5) $\qquad A^e = \begin{bmatrix} A+BKC & BL \\ MC & N \end{bmatrix}$

Theorem VI.8.

The following statements are equivalent:

i)   The system (6.1) is stabilizable-detectable.

ii)   The exists a compensator (N,M,L,K) with finite-dimensional state space and $K=0$ such that the closed loop system (6.4) is stable.

Proof:

See Jacobson and Nett [21]. $\qquad\qquad\qquad\qquad\qquad\qquad\qquad\qquad\qquad$ $\square$

Now we shall combine the notion of invariance with stability as in Curtain [8] and Hautus [19].

Definition VI.9: Stabilizability Subspace.

A subspace V of $\mathcal{H}$ is stabilizability subspace if there exists a bounded feedback law F, such that

a) $T_{A+BF}(t)V \subset V$ and

b) $\exists \delta > 0$ such that $\|T_{A+BF}(t)|_V\| \leq Me^{-\delta t} \qquad \forall t \geq 0$

Remark:

In Basile, Marro and Piazzi [2] this notion was called internally stable.

Combining the equivalence between open loop, frequency domain and closed loop invariance with theorem VI.3. gives

Theorem VI.10.

Let V be a closed subspace of $\mathcal{H}$. Then the following assertions are equivalent.

i) V is a stabilizability subspace.

ii)   For every x_0 in V there exists a $u(t) \in C([0,\infty);\mathbb{R}^m) \cap L^2([0,\infty);\mathbb{R}^m)$

such that

$$(6.6) \qquad x(t) = T_A(t)x_0 + \int_0^t T_A(t-s)Bu(s)ds$$

is contained in V and $x(.) \in L^2([0,\infty);V)$.

iii) For every $x_0 \in V$ there exists a $\xi(.) \in H^2(V)$ and $\omega(.) \in H^2(\mathbf{R}^m)$ such that $\lim_{\substack{s \to \infty \\ s \in R}} s\omega(s)$ exists, $\xi(s) \in V \cap D(A)$ and

$$(6.7) \qquad x_0 = (s-A)\xi(s) - B\omega(s); \quad s \in \mathbf{C}_+ : = \{s \in \mathbf{C} \mid Re(s) > 0\}.$$

Proof:

i)$\Rightarrow$ii):

From definition VI.9 we have the existence of a bounded feedback law F such that the closed loop system satisfies $T_{A+BF}(t)V \subset V$ and $A+BF|_V$ is stable. From Curtain & Pritchard theorem 2.31 we have that

$$(6.8) \qquad T_{A+BF}(t)x_0 = T_A(t)x_0 + \int_0^t T_A(t-s)BFT_{A+BF}(s)x_0 ds$$

Defining $x(t) = T_{A+BF}(t)x_0$ and $u(t) = FT_{A+BF}(t)x_0$ gives the desired result.

ii)$\Rightarrow$iii):

Taking the Laplace transform of (6.6) gives

$$(6.9) \qquad \xi(s) = (s-A)^{-1}x_0 + (s-A)^{-1}B\omega(s)$$

where $\xi(s)$ and $\omega(s)$ are respectively the Laplace transform of $x(.)$ and of $u(.)$. Since $x(.)$ and $u(.)$ are square integrable, $\xi(.)$ and $\omega(.)$ are H^2 functions. By the definition of the Laplace transform we have that $\xi(s) \in V$ and equation (6.9) gives that $\xi(s) \in D(A)$. Rewriting equation (6.9) gives

$$(6.10) \qquad x_0 = (s-A)\xi(s) - B\omega(s)$$

and since $\xi(.)$ and $\omega(.)$ are H^2 functions we have that (6.10) holds on $\mathbf{C}_+$. The strictly properness of $\omega(.)$ follows from the fact that $u(.)$ is continuous, Doetsch [16, p. 226].

iii)→i):

From theorem II.27 we have the existence of a bounded feedback law F_1 such that $T_{A+BF_1}(t)V \subset V$. So we have the invariance, but no further stability properties yet.

From equation (6.7) we have that

(6.11) $$x_0 = (s - A - BF_1)\xi(s) - B\Big\{\omega(s) - F_1\xi(s)\Big\}.$$

By the invariance of V under $T_{A+BF_1}(t)$ we have that $(s - A - BF_1)(V \cap D(A)) \subset V$. So $B\Big\{\omega(s) - F_1\xi(s)\Big\} \in V$. Let $\mathcal{U}_1$ denote the subset of $\mathbf{R}^m$ such that $B(\mathcal{U}_1) = P_V \operatorname{Im} B$.

Then by definition of $\mathcal{U}_1$ we have that $\Big\{\omega(s) - F_1\xi(s)\Big\} \in \mathcal{U}_1$.

Consider now the system $\{(A+BF_1), B\}$ with state space V and input space $\mathcal{U}_1$. Theorem VI.3.*iv)* with (6.11) implies that this system is stabilizable. Thus there exists a bounded feedback law $F_2: V \mapsto \mathcal{U}_1$ such that $(A + BF_1 + BF_2)$ generates a stable semigroup on V. Now we have to construct a feedback law on the whole state space such that V is a stabilizability subspace. Define F as

$$F|_V = F_1 + F_2 \; and$$
$$F|_{V^\perp} = F_1$$

Then $\big(BF - BF_1\big)x = BF_2 P_V x \in V$, since $F_2(V) \subset \mathcal{U}_1$. So V is $T_{A+BF}(t)$ invariant, furthermore $BF(V) = \big(BF_1 + BF_2\big)(V)$. So with lemma V.1 we have that $T_{A+BF}(t)|_V = T_{A+BF_1+BF_2}(t)|_V$ and so V is stabilizability subspace. □

The next rather technical lemma will show the usefulness of stabilizability subspaces in relation with maintaining stabilizability under finite dimensional extensions. We shall need this for the disturbance decoupling problem.

Lemma VI.11.

Let (A, B) be stabilizable, and let V be a stabilizability subspace. For a space W with $\dim(W) < \infty$ and $R \in \mathcal{L}(\mathcal{H}, W)$ we make the following extensions of A, B and V:

$$\tilde{A} = \begin{pmatrix} A & 0 \\ 0 & 0 \end{pmatrix}, \quad D(\tilde{A}) = D(A) \oplus W; \quad \tilde{B} = \begin{pmatrix} B & 0 \\ 0 & I \end{pmatrix} \quad and \quad V^e = \left\{ \begin{pmatrix} x \\ Rx \end{pmatrix} \Big| x \in V \right\}$$

If V^e is closed loop invariant for the system $(\tilde{A},\tilde{B})$, then the system $(\tilde{A}+\tilde{B}\tilde{F}|_{V^e},\tilde{B})$ with state space V^e and input space $\mathcal{U}_1 := \tilde{B}^{-1}(V^e)$ is stabilizable, where $\tilde{F}$ is any feedback law that satisfies: $T_{\tilde{A}+\tilde{B}\tilde{F}}(t)V^e \subset V^e$.

Proof:

Since (A,B) is stabilizable, we have from the nice form of $\tilde{A}$ and $\tilde{B}$ that $(\tilde{A},\tilde{B})$ is stabilizable too. Furthermore since stabilizability is not effected by feedback we have that the system $(\tilde{A}+\tilde{B}\tilde{F},\tilde{B})$ is stabilizable. Thus we have from lemma VI.5 that V^e can be decomposed as $V^e = V_u^e \oplus V_s^e$ where $\dim(V_u^e) < \infty$, V_u^e and V_s^e are both semigroup $(\tilde{A}+\tilde{B}\tilde{F})$ invariant and $\tilde{A}+\tilde{B}\tilde{F}|_{V_s^e}$ is stable. We shall begin by proving that the finite dimensional system $(\tilde{A}+\tilde{B}\tilde{F}|_{V_u^e}, P_{V_u^e}\tilde{B})$ is controllable.

Let $F \in \mathcal{L}(\mathcal{H};\mathbb{R}^m)$ be such that $T_{A+BF}(t)V \subset V$ and $\|T_{A+BF}(t)|_V\| \le Me^{-\delta t}$, then for x in V, $(s-A-BF)^{-1}x \in V$ and this is an analytic function on $\mathbb{C}_{-\delta,+} = \{s \in \mathbb{C}\,|\,Re\,s > -\delta\}$. Define $\xi(s) = (s-A-BF)^{-1}x \in V$; then if $\tilde{F} = \begin{bmatrix} \tilde{F}_{11} & \tilde{F}_{12} \\ \tilde{F}_{21} & \tilde{F}_{22} \end{bmatrix}$, we have

$$\begin{bmatrix} x \\ Rx \end{bmatrix} = \begin{bmatrix} s-A-BF & 0 \\ 0 & s \end{bmatrix}\begin{bmatrix} \xi(s) \\ R\xi(s) \end{bmatrix} + \begin{bmatrix} 0 \\ -sR\xi(s)+Rx \end{bmatrix} =$$

(6.12)

$$= \begin{bmatrix} s-A-B\tilde{F}_{11} & -B\tilde{F}_{12} \\ -\tilde{F}_{21} & s-\tilde{F}_{22} \end{bmatrix}\begin{bmatrix} \xi(s) \\ R\xi(s) \end{bmatrix} + \begin{bmatrix} B & 0 \\ 0 & I \end{bmatrix}\begin{bmatrix} (\tilde{F}_{11}-F)\xi(s)+\tilde{F}_{12}R\xi(s) \\ \tilde{F}_{21}\xi(s)+\tilde{F}_{22}R\xi(s)-sR\xi(s)+Rx \end{bmatrix}$$

Since V^e is $(s-\tilde{A}-\tilde{B}\tilde{F})$ invariant we have that

$$\begin{bmatrix} B & 0 \\ 0 & I \end{bmatrix}\begin{bmatrix} (\tilde{F}_{11}-F)\xi(s)+\tilde{F}_{12}R\xi(s) \\ \tilde{F}_{21}\xi(s)+\tilde{F}_{22}R\xi(s)-sR\xi(s)+Rx \end{bmatrix} \in V^e$$

and thus

$$\begin{bmatrix} (\tilde{F}_{11}-F)\xi(s)+\tilde{F}_{12}R\xi(s) \\ \tilde{F}_{21}\xi(s)+\tilde{F}_{22}R\xi(s)-sR\xi(s)+Rx \end{bmatrix} \in \mathcal{U}_1.$$

So every $\begin{bmatrix} x \\ Rx \end{bmatrix}$ in V^e can be written as

(6.13)

$$\begin{bmatrix} x \\ Rx \end{bmatrix} = (s-\tilde{A}-\tilde{B}\tilde{F})|_{V^e}\xi^e(s)+\tilde{B}\,w^e(s)$$

where $\xi^e(s) = \begin{bmatrix} \xi(s) \\ R\xi(s) \end{bmatrix}$ and $w^e(s) \in \mathcal{U}_1$ are analytic on $\mathbb{C}_{-\delta,+}$. Let $\begin{bmatrix} x \\ Rx \end{bmatrix}$ be in V_u^e, the finite dimensional unstable part of $\tilde{A}+\tilde{B}\tilde{F}|_{V^e}$, then

$$\begin{pmatrix} x \\ Rx \end{pmatrix} = P_{V_u^e}\begin{pmatrix} x \\ Rx \end{pmatrix} = P_{V_u^e}(s - \tilde{A} - \tilde{B}\tilde{F})|_{V^e}\xi^e(s) + P_{V_u^e}\tilde{B}\,\omega^e(s) \; or$$

(6.14) $\qquad \begin{pmatrix} x \\ Rx \end{pmatrix} = (s - \tilde{A} - \tilde{B}\tilde{F})|_{V_u^e}\xi^e(s) + P_{V_u^e}\tilde{B}\,\omega^e(s)$

So to every $\begin{pmatrix} x \\ Rx \end{pmatrix} \in V_u^e$ there exists a pair $(\xi^e(.),\omega^e(.))$ which are analytic on $\mathbb{C}_{-\delta,+}$. Since the spectrum of $(\tilde{A}+\tilde{B}\tilde{F}|_{V_u^e})$ is contained in $\mathbb{C}_{-\delta,+}$ we have from the well-known Hautus test that equation (6.14) shows that $(\tilde{A}-\tilde{B}\tilde{F}|_{V_u^e}, P_{V_u^e}\tilde{B})$ is controllable.

Now we can consider the system $(\tilde{A}-\tilde{B}\tilde{F}|_{V^e}, P_{V^e}\tilde{B})$ with state space V^e and input space $\mathcal{U}_1$. Since we can decompose this state space into a stable part and a finite dimensional controllable part we have from theorem VI.3. that this system is stabilizable.

$\hfill \square$

With the concept of stabilizability subspace we can introduce the notion of a stable (C,A,B)-pair.

Definition VI.12: Stable (C,A,B)-pair

A (C,A,B)-pair (S,V) is a stable (C,A,B)-pair if there exists $(F,G)\in \mathcal{L}(\mathcal{H};\mathbb{R}^m) \oplus \mathcal{L}(\mathbb{R}^k;\mathcal{H})$ such that

a) $\qquad T_{A+BF}(t)V \subset V, \quad T_{A+GC}(t)S \subset S \;$ and

b) $\qquad \exists \delta > 0, \; M \geq 1 \;$ such that

(6.15) $\qquad \|T_{A+BF}(t)|_V\| \leq Me^{-\delta t}$

(6.16) $\qquad \|T_{A+GC}(t) \, mod \, S\| \leq Me^{-\delta t}$

Since there is a close relationship between modulo spaces and the orthogonal complement, we have the following result.

Lemma VI.13.

Let $Q \in \mathcal{L}(\mathcal{H})$ and let V be a closed invariant subspace of Q, then

$$\|Q \, mod \, V\| = \|P_{V^\perp}QP_{V^\perp}\| = \|Q^*|_{V^\perp}\|$$

Proof:

See Kato [22] and Yosida [44].

With this lemma we have

Lemma VI.14.

A pair (S,V) of closed subspaces is a stable (C,A,B)–pair if and only
if

i) $S \subset V$

ii) V is a stabilizability subspace for the system (A,B)

iii) $S^\perp$ is a stabilizability subspace for the system (A^*,C^*).

Proof:

This follows easily from definition VI.9 and lemma VI.13.

$\square$

**Section VI.2: Disturbance Decoupling Problem with Measurement Feedback
and Stability**

In this section we shall give necessary and sufficient conditions for
the solvability of DDPMS. As with DDP and DDPM these conditions will be
stated in terms of the existence of certain subspaces.

Consider the system on the Hilbert space $\mathcal{H}$.

$$\dot{x} = Ax + Bu + Eq$$

(6.17)

$$y = Cx, \quad z = Dx$$

where A, B, C, D, E are the same as in section V.2, q is the disturbance
and z is the output to be decoupled. The DDPMS is to design a dynamic
output feedback of the form (6.4) such that, in closed loop, the output
$z(t)$ is decoupled from the disturbance $q(t)$. i.e.

(6.18) $$D^e \int_0^t T_{A^e}(t-s)E^e q(s)ds = 0, \quad \forall t \geq 0$$

and $T_{A^e}(t)$ is stable, i.e. there exists M, $\delta > 0$, such that

(6.19) $\|T_{A^e}(t)\| \le M e^{-\delta t}$

where $D^e = (D,0)$, $E^e = \begin{bmatrix} E \\ 0 \end{bmatrix}$ and $T_{A^e}(t)$ is the semigroup generated by the closed loop system operator A^e (6.5).

Pictorially our configuration is given in figure 6.1.

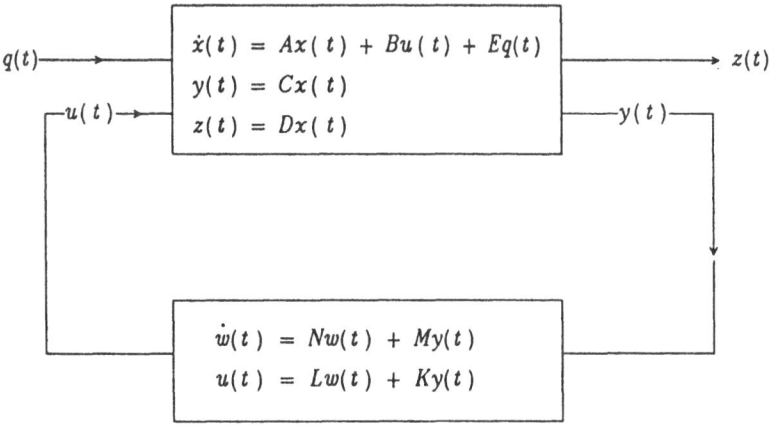

$$\dot{x}(t) = Ax(t) + Bu(t) + Eq(t)$$
$$y(t) = Cx(t)$$
$$z(t) = Dx(t)$$

$$\dot{w}(t) = Nw(t) + My(t)$$
$$u(t) = Lw(t) + Ky(t)$$

figure 6.1

Now we can state our main result which is completely similar to the finite dimensional case. For finite dimensional systems the condition $dim(V/S) < \infty$ is trivially satisfied, and thus is omited there.

Theorem VI.15.

DDPMS is solvable with a finite dimensional compensator if and only if

i) (A,B) is stabilizable

ii) (C,A) is detectable

iii)   There exists a stable (C,A,B)–pair, (S,V) such that $dim(V/S) < \infty$ and

$$Im\ E \subset S \subset V \subset Ker\ D$$

Proof:

only if)

This proof is essentially the same as the only if part of theorem V.10. So suppose that DDPMS is solvable. Thus there exists a finite dimensional compensator such that $D^e T_{A^e}(t)E^e \equiv 0$ and $T_{A^e}(t)$ is stable.

Define V^e to be the closed span of the space consisting of all x^e such that

$$x^e = \int_0^t T_{A^e}(t-s)E^e q(s)ds \text{ for some } t \text{ and } q(s)$$

This subspace is $T_{A^e}(t)$–invariant and by lemma V.6. the pair (S,V) satisfies $S \subset V$ and $dim(V/S) < \infty$, where

$$S = \{x \in \mathcal{H} \mid \begin{bmatrix} x \\ 0 \end{bmatrix} \in V^e\}$$

$$V = \{x \in \mathcal{H} \mid \begin{bmatrix} x \\ w \end{bmatrix} \in V^e \text{ for some } w \in W\}.$$

We shall prove that (S,V) is a stable (C,A,B)–pair. Let x be an element of V, then there exists a $w \in W$ such that $\begin{bmatrix} x \\ w \end{bmatrix}$ is an element of V^e. From theorem 2.31 of Curtain and Pritchard [9] we have that

$$T_{A^e}(t)\begin{bmatrix} x \\ w \end{bmatrix} = \begin{bmatrix} T_A(t) & 0 \\ 0 & T_N(t) \end{bmatrix}\begin{bmatrix} x \\ w \end{bmatrix} +$$

$$\int_0^t \begin{bmatrix} T_A(t-s) & 0 \\ 0 & T_N(t-s) \end{bmatrix}\begin{bmatrix} BKC & BL \\ MC & N \end{bmatrix}\begin{bmatrix} T_{A^e}(s)\begin{bmatrix} x \\ w \end{bmatrix} \end{bmatrix}ds$$

So

$$P_{\mathcal{H}}T_{A^e}(t)x = T_A(t)x + \int_0^t T_A(t-s)B\{KCP_{\mathcal{H}} + LP_W\}T_{A^e}(s)\begin{bmatrix} x \\ w \end{bmatrix}ds.$$

Since $T^e(t)\begin{bmatrix} x \\ w \end{bmatrix} \in V^e$, we have that $P_{\mathcal{H}}T^e(t)x \in V$ and so for every x in V, there exists a $u(t) = \{KCP_{\mathcal{H}} + LP_W\}T_{A^e}(t)\begin{bmatrix} x \\ w \end{bmatrix} \in C([0,\infty);\mathbb{R}^m) \cap L^2([0,\infty);\mathbb{R}^m)$ such that

$$x(t) = T_A(t)x + \int_0^t T_A(t-s)Bu(s)ds$$

is in V and $x(t) \in L^2([0,\infty);V)$. So V is a stabilizability subspace. With the same reasoning one can prove that $S^{\perp}$ is a stabilizability subspace for the system (A^*,C^*). So (S,V) is a stable (C,A,B)–pair with $dim(V/S) < \infty$ and by the definition of S and V we have that $Im\,E \subset S \subset V \subset Ker\,D$. The conditions $i)$ and $ii)$ are a direct consequence of theorem VI.8.

(if)

In this sufficiency part we shall follow closely the line of Willems and Commault [41].

Step 1. Disturbance decoupling

From theorem V.12. we have the existence of a finite dimensional compensator with state space $W_1 \cong V/S$ and a closed subspace $V_1^e = \left\{ \begin{bmatrix} x \\ Rx \end{bmatrix} \middle| x \in V \right\}$; $Im\, R = W_1$; $Ker\, R = S$ which is $T_{A^e,1}(t)$–invariant, where $T_{A^e,1}(t)$ is the semigroup of the closed loop system $A^{e,1} = \begin{bmatrix} A+BK^1C & BL^1 \\ M^1C & N^1 \end{bmatrix}$. Furthermore

(6.20) $\qquad Im\, E^{e,1} \subset V_1^e \subset Ker\, D^{e,1}$

where $E^{e,1} = \begin{bmatrix} E \\ 0 \end{bmatrix}$ and $D^{e,1} = (D,0)$.

So the system is disturbance decoupled but enjoys no further stability properties as yet. Later on in this proof we shall solve the DDPMS. We shall consider the extended system $(A^{e,1}, \tilde{B}, \tilde{C}, D^{e,1}, E^{e,1})$ on the state space $\mathcal{H}^{e,1}$, where

$$\mathcal{H}^{e,1} = \mathcal{H} \oplus W^1;$$

$$A^{e,1}: \mathcal{H}^{e,1} \to \mathcal{H}^{e,1}, \quad A^{e,1} = \begin{bmatrix} A+BK^1C & BL^1 \\ M^1C & N^1 \end{bmatrix}; \quad D(A^e,{}^1) = D(A) \oplus W^1;$$

$$\tilde{B}: \mathbb{R}^m \oplus W^1 \to \mathcal{H}^{e,1}, \quad \tilde{B} = \begin{bmatrix} B & 0 \\ 0 & I \end{bmatrix} \text{ and}$$

$$\tilde{C}: \mathcal{H}^{e,1} \to \mathbb{R}^p \oplus W^1, \quad \tilde{C} = \begin{bmatrix} C & 0 \\ 0 & I \end{bmatrix},$$

Thus the system $(A^{e,1}, \tilde{B}, \tilde{C})$ is the closed loop system from step 1 with extra inputs and outputs. In formula this system is given by

$$\dot{x}^{e,1} = A^{e,1} x^{e,1} + \tilde{B} u^{e,1} + E^{e,1} q$$

(6.21)

$$y^{e,1} = \tilde{C} x^{e,1}, \quad z = D^{e,1} x^{e,1}$$

where $x^{e,1} = \begin{bmatrix} x \\ w_1 \end{bmatrix}$, $u^{e,1} = \begin{bmatrix} u \\ w \end{bmatrix}$ and $y^{e,1} = \begin{bmatrix} y \\ w_1 \end{bmatrix}$.

In signal flow graph the system is given by

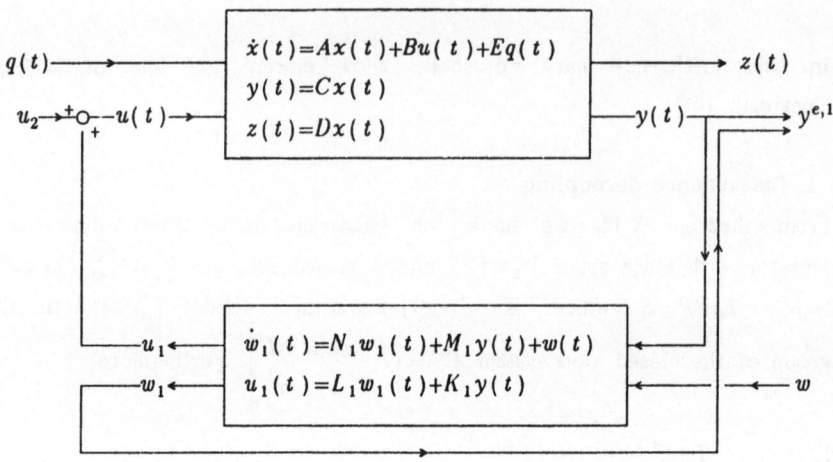

figure 6.2.

Then since W^1 is finite dimensional, we have that the system $(\tilde{C}, A^{e,1}, \tilde{B})$ satisfies the same conditions as the system (C, A, B) and furthermore $(A^{e,1}, \tilde{B})$ is stabilizable and $(\tilde{C}, A^{e,1})$ is detectable.

Later on in the proof we shall construct a stabilizing compensator for the system $(A^{e,1}, \tilde{B}, \tilde{C})$ under which the disturbance decoupling is maintained. This compensator is based on the following decomposition of the state space, input space and output space of the system $(\tilde{C}, A^{e,1}, \tilde{B})$

$$
(6.22)\quad
\begin{cases}
\mathcal{H}^{e,1} = V_1^e \oplus V_1^{e\perp} \\[2mm]
\mathbf{R}^m \oplus W^1 = \tilde{B}^{-1} V_1^e \oplus \tilde{B}^*\left[V_1^{e\perp}\right], \quad \text{where } \tilde{B}^{-1} V_1^e \text{ is the set of all} \\
\hphantom{\mathbf{R}^m \oplus W^1 = \tilde{B}^{-1} V_1^e \oplus \tilde{B}^*\left[V_1^{e\perp}\right],\quad} \text{elements } u \text{ in } \mathbf{R}^m \oplus W^1 \text{ such that } \tilde{B}u \in V_1^e \\[2mm]
\mathbf{R}^p \oplus W^1 = \tilde{C} V_1^e \oplus \tilde{C}^{*-1}\left[V_1^{e\perp}\right]
\end{cases}
$$

We remark that the decomposition of the input space and output space holds

since $\tilde{B}^{-1} V_1^e = \left[\tilde{B}^*\left[V_1^{e\perp}\right]\right]^{\perp}$ and $\left[\tilde{C} V_1^e\right]^{\perp} = \tilde{C}^{*-1}\left[V_1^{e\perp}\right]$. Furthermore we define

$$
\mathcal{U}_1 = \tilde{B}^{-1} V_1^e, \quad \mathcal{U}_2 = \tilde{B}^*\left[V_1^{e\perp}\right], \quad \mathcal{Y}_1 = \tilde{C} V_1^e \text{ and } \mathcal{Y}_2 = \tilde{C}^{*-1}\left[V_1^{e\perp}\right]
$$

In this basis we can decompose the system $(A^{e,1}, \tilde{B}, \tilde{C}, D^{e,1}, E^{e,1})$ as follows.

$$(6.23) \begin{cases} A_{11}^{e,1}: V_1^e \cap D(A) \mapsto V_1^e; \quad A_{11}^{e,1} = A^{e,1}|_{V_1^e}, \\[2mm] \left[A_{22}^{e,1}\right]^*: V_1^{e^\perp} \cap D(A^*) \mapsto V_1^{e^\perp}; \quad \left[A_{22}^{e,1}\right]^* = A^{e,1*}|_{V_1^{e^\perp}} \\[2mm] A_{22}^{e,1}: V_1^{e^\perp} \cap D(A_{22}^{e,1}) \mapsto V_1^{e^\perp}; \qquad A_{22}^{e,1} \text{ is the dual of } \left[A_{22}^{e,1}\right]^* \text{ in the} \\[2mm] \qquad\qquad\qquad\qquad\qquad\qquad\qquad\qquad \text{space } V_1^{e^\perp} \\[2mm] \tilde{B}_{11}: \mathcal{U}_1 \mapsto V_1^e; \quad \tilde{B}_{11}u = \tilde{B}u \\[2mm] \tilde{B}_{12}: \mathcal{U}_2 \mapsto V_1^e; \quad \tilde{B}_{12} = P_{V_1^e}\tilde{B}|_{\mathcal{U}_2} \\[2mm] \tilde{B}_{22}: \mathcal{U}_2 \mapsto V_1^{e^\perp}; \quad \tilde{B}_{12} = P_{V_1^{e^\perp}}\tilde{B}|_{\mathcal{U}_2} \\[2mm] \tilde{C}_{11}: V_1^e \mapsto \mathcal{Y}_1; \quad \tilde{C}_{11} = \tilde{C}|_{V_1^e} \\[2mm] \tilde{C}_{12}: V_1^{e^\perp} \mapsto \mathcal{Y}_1; \quad \tilde{C}_{12} = P_{\mathcal{Y}_1}\tilde{C}|_{V_1^{e^\perp}} \\[2mm] \tilde{C}_{22}: V_1^{e^\perp} \mapsto \mathcal{Y}_2; \quad \tilde{C}_{22} = P_{\mathcal{Y}_2}\tilde{C}|_{V_1^{e^\perp}} \end{cases}$$

We remark that $A_{11}^{e,1}$, $\left[A_{22}^{e,1}\right]^*$ are the generators of $T_{A^e,1}(t)|_{V_1^e}$ and $T_{A^e,1}^*(t)|_{V_1^{e^\perp}}$ respectively.

So heuristically we have decomposed the system (6.21) into

$$\begin{pmatrix} \dot{x}_1 \\ \dot{x}_2 \end{pmatrix} = \begin{pmatrix} A_{11}^{e,1} & A_{12}^{e,1} \\ 0 & A_{22}^{e,1} \end{pmatrix} \begin{pmatrix} x_1 \\ x_2 \end{pmatrix} + \begin{pmatrix} \tilde{B}_{11} & \tilde{B}_{12} \\ 0 & \tilde{B}_{22} \end{pmatrix} \begin{pmatrix} u_1 \\ u_2 \end{pmatrix} + \begin{pmatrix} E^{e,1} \\ 0 \end{pmatrix} q$$

(6.24)

$$\begin{pmatrix} y_1 \\ y_2 \end{pmatrix} = \begin{pmatrix} \tilde{C}_{11} & \tilde{C}_{12} \\ 0 & \tilde{C}_{22} \end{pmatrix} \begin{pmatrix} x_1 \\ x_2 \end{pmatrix}; \quad z = \left(0, D^{e,1}\right) \begin{pmatrix} x_1 \\ x_2 \end{pmatrix}.$$

where $A_{12}^{e,1} = "P_{V_1^e}A^{e,1}|_{V_1^{e^\perp}}"$, and the zero in the $A^{e,1}$ is caused by the invariance of V_1^e, the zero in $\tilde{B}$ and $\tilde{C}$ is a consequence of the decomposition of the input and output space and the zero in $D^{e,1}$ and $E^{e,1}$ is a consequence of (6.20). The signal flow graph of (6.24) may be visualized as shown in figure 6.3.

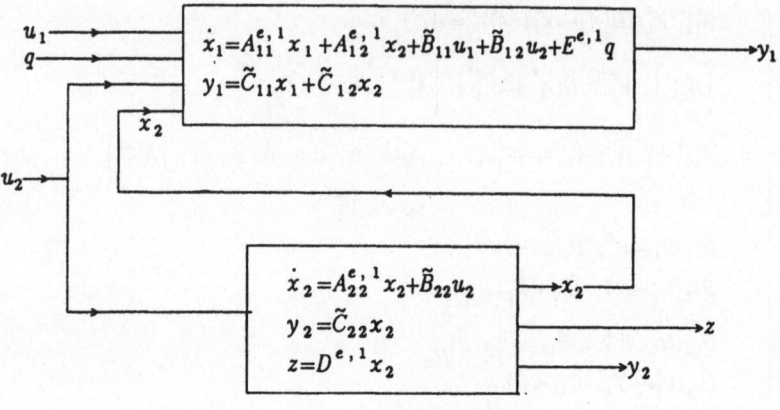

$$\dot{x}_1 = A_{11}^{e,1} x_1 + A_{12}^{e,1} x_2 + \tilde{B}_{11} u_1 + \tilde{B}_{12} u_2 + E^{e,1} q$$
$$y_1 = \tilde{C}_{11} x_1 + \tilde{C}_{12} x_2$$

$$\dot{x}_2 = A_{22}^{e,1} x_2 + \tilde{B}_{22} u_2$$
$$y_2 = \tilde{C}_{22} x_2$$
$$z = D^{e,1} x_2$$

figure 6.3

In step 2 we shall prove that the system $(A_{11}^{e,1}, \tilde{B}_{11}, \tilde{C}_{11})$ is stabilizable and detectable, and in step 3 we shall prove the same for $(A_{22}^{e,1}, \tilde{B}_{22}, \tilde{C}_{22})$. Let $(N_2, M_2, L_2, 0)$ and $(N_3, M_3, L_3, 0)$ denote the stabilizing compensator for respectively $(A_{11}^{e,1}, \tilde{B}_{11}, \tilde{C}_{11})$ and $(A_{22}^{e,1}, \tilde{B}_{22}, \tilde{C}_{22})$. Then in closed loop (figure 6.4) the state $\begin{bmatrix} x_1 \\ x_2 \end{bmatrix}$ is exponentially decaying. In step 4 we shall present a precise proof. Now we shall only give an intuitively one. The decay of x_2 is obvious since the compensator $(N_3, M_3, L_3, 0)$ stabilised the system $(A_{22}^{e,1}, \tilde{B}_{22}, \tilde{C}_{22})$ and this system is not influenced by other inputs. Furthermore we have that u_2 is exponentially decaying. Now in closed loop the upper part of figure 6.4 is a stable system with exponentially decaying inputs, and thus x_1 is exponentially decaying.

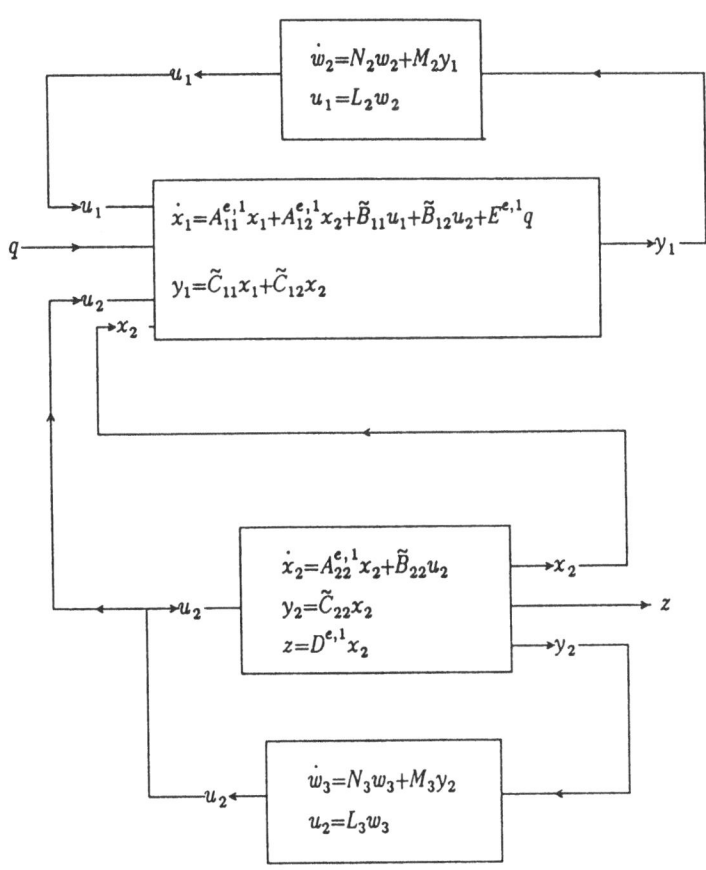

figure 6.4

Step 2: Disturbance loop stabilization

We shall prove that the system $(A_{11}^{e,1}, \tilde{B}_{11}, \tilde{C}_{11})$, with state space V_1^e, input space U_1 and output space Y_1 is stabilizable–detectable.

The stabilizability of the system $(A_{11}^{e,1}, \tilde{B}_{11})$ is a direct consequence of lemma VI.11.

The detectability of $(\tilde{C}_{11}, A_{11}^{e,1})$ follows easily from the detectability of $(\tilde{C}, \tilde{A})$ and is left to the reader.

Now from theorem VI.8 we have that there exists a finite dimensional compensator with state space W^2 and no feed–through term such that

$$\begin{pmatrix} A_{11}^{e,1} & \tilde{B}_{11}L_2 \\ M_2\tilde{C}_{11} & N_2 \end{pmatrix}$$

is stable on the state space $V_1^e \oplus W_2$.

Step 3: Controlled output stabilization

With a similar argument as used in step 2, one can prove that the system $(\tilde{B}_{22}^{*}, (A_{22}^{e;1})^{*}, \tilde{C}_{22}^{*})$ with state space $V_1^{e\perp}$, input space $\mathcal{Y}_2$ and output space $\mathcal{U}_2$ is stabilizable/detectable. So there exists a compensator with state space W_3, $dim(W_3) < \infty$ such that $\begin{pmatrix} (A_{22}^{e;1})^{*} & \tilde{C}_{22}^{*} & M_3^{*} \\ L_3^{*}\tilde{B}_{22}^{*} & N_3^{*} \end{pmatrix}$ is stable on the state space $V_1^{e\perp} \oplus W_3$. So the dual is also stable, this dual is given by

$$\begin{pmatrix} A_{22}^{e;1} & \tilde{B}_{22}L_3 \\ M_3\tilde{C}_{22} & N_3 \end{pmatrix}.$$

Step 4: DDPMS

It remains to show that the connection of the three constructed compensators solves the DDPMS.

Applying both compensators to the system (6.21) gives for the closed loop system the following system operator

$$(6.25) \qquad A^e = \begin{pmatrix} A^{e,1} & \tilde{B}\,i_{U_1}L_2 & \tilde{B}\,i_{U_2}L_3 \\ M_2 P_{Y_1}\tilde{C} & N_2 & 0 \\ M_3 P_{Y_2}\tilde{C} & 0 & N_3 \end{pmatrix}$$

with $D(A^e) = D(A) \oplus W_1 \oplus W_2 \oplus W_3$. and i denotes the inclusion operator. Let $T_{A^e}(t)$ denote the semigroup generated by A^e and define V^e as $V^e := V_1^e \oplus W_2 \oplus \{0\}$. We shall prove that V^e is $T_{A^e}(t)$-invariant. This follow easily from the fact that V^e is invariant for the semigroup generated by $\begin{pmatrix} A^{e,1} & 0 & 0 \\ 0 & N_2 & 0 \\ 0 & 0 & N_3 \end{pmatrix}$ and furthermore from the definition of $\mathcal{U}_1$ and $\mathcal{Y}_2$ we have that

$$(6.26) \qquad \begin{pmatrix} 0 & \tilde{B}\,i_{U_1}L_2 & \tilde{B}\,i_{U_2}L_3 \\ M_2 P_{Y_1}\tilde{C} & 0 & 0 \\ M_3 P_{Y_2}\tilde{C} & 0 & 0 \end{pmatrix} \begin{pmatrix} V_1^e \\ \oplus \\ W_2 \\ \oplus \\ 0 \end{pmatrix} \subset \begin{pmatrix} V_1^e \\ \oplus \\ W_2 \\ \oplus \\ 0 \end{pmatrix}$$

and so lemma II.2 gives the desired result. Furthermore since $E^e := \begin{pmatrix} E \\ 0 \\ 0 \\ 0 \end{pmatrix}$ and $D^e = (D,0,0,0)$, we have $Im\,E^e \subset V^e \subset Ker\,D^e$ and thus we have still disturbance decoupling. It remains to show that A^e is stable.

From the definition of A^e and lemma V.1. we have that $T_{A^e}(t)$ restricted

to V^e is the same as the semigroup generated by $\begin{bmatrix} A_{11}^{e,1} & \tilde{B}_{11}L_2 \\ M_2\tilde{C}_{11} & N_2 \end{bmatrix}$, which we

shall denote by $T_{A^e,2}(t)$. From the disturbance loop stabilization we have that this is stable.

In the same way one can prove that $T_{A^e}(t)^*$ restricted to $V^{e\perp}$ is the same as

$T_{A^e,3}(t)^*$, the semigroup generated by $\begin{bmatrix} (A_{22}^{e,1})^* & \tilde{C}_{22}^* & M_3^* \\ L_3^*\tilde{B}_{22}^* & N_3^* \end{bmatrix}$, which was also

stable.

With these stable semigroups we can show that $T_{A^e}(t)$ is stable. Since (A,B) is stabilizable and W_1, W_2 and W_3 are finite dimensional the system $(\begin{bmatrix} A & 0 & 0 & 0 \\ 0 & 0 & 0 & 0 \\ 0 & 0 & 0 & 0 \\ 0 & 0 & 0 & 0 \end{bmatrix}, \begin{bmatrix} B & 0 & 0 & 0 \\ 0 & I & 0 & 0 \\ 0 & 0 & I & 0 \\ 0 & 0 & 0 & I \end{bmatrix})$ is stabilizable too, and since feedback does not

affect stabilizability we have that $(A^e, \begin{bmatrix} B & 0 & 0 & 0 \\ 0 & I & 0 & 0 \\ 0 & 0 & I & 0 \\ 0 & 0 & 0 & I \end{bmatrix})$ is stabilizable. Thus in

order to prove the stability of A^e it is sufficient to prove that A^e has no unstable eigenvalues.

Suppose that $T_{A^e,2}(t)$ and $T_{A^e,3}(t)$ satisfies respectively $\|T_{A^e,2}(t)\| < M_2 e^{-\delta_2 t}$ and $\|T_{A^e,3}(t)\| < M_3 e^{-\delta_3 t}$ for some δ_2 and δ_3 larger than zero, note that these constants exist since both semigroups are stable. Let $\lambda \in \mathbb{C}$ with $Re\lambda > max(-\delta_2, -\delta_3)$ be an unstable eigenvalue, then there exists a $x^e \in \mathcal{H}^e = \mathcal{H} \oplus W_1 \oplus W_2 \oplus W_3$ such that

$$T_{A^e}(t)x^e = e^{\lambda t}x^e \quad \Rightarrow P_{V^e\perp}T_{A^e}(t)x^e = e^{\lambda t}P_{V^e\perp}x^e,$$

since V^e is $T_{A^e}(t)$–invariant, we have that

$$P_{V^e\perp}T_{A^e}(t)P_{V^e\perp}x^e = e^{\lambda t}P_{V^e\perp}x^e$$

Thus if $P_{V^e\perp}x^e \neq 0$, then $e^{\lambda t}$ is an element of the point spectrum of $P_{V^e\perp}T_{A^e}(t)P_{V^e\perp}$, and thus $e^{\lambda t}$ is in the point spectrum of $P_{V^e\perp}(T_{A^e}(t))^*P_{V^e\perp}$ $= (T_{A^e}(t))^*|_{V^e\perp} = (T_{A^e,3}(t))^*$. By the stability of $T_{A^e,3}(t)^*$ this is not true, so $P_{V^e\perp}x^e = 0$, or equivalently $x^e \in V^e$. However for all $t \geq 0$ we have that

$$\|e^{\lambda t}x^e\| = \|T_{A^e}(t)x^e\| = \|T_{A^e}(t)|_{V^e}x^e\| = \|T_{A^e,2}(t)x^e\| \leq M_2 e^{-\delta_2 t}\|x^e\|$$

So $x^e \equiv 0$, which implies that $\sigma(A^e) \subset \mathbb{C}_-$ and corollary VI.4. gives that A^e is stable.

Thus the DDPMS is solved with the compensator (N^e, M^e, L^e, K^e) with finite dimensional state space $W^e = W_1 \oplus W_2 \oplus W_3$ and

$$
(6.27) \quad
\begin{cases}
N^e = \begin{pmatrix} N_1 & P_{W_1} i_{U_1} L_2 & P_{W_1} i_{U_2} L_3 \\ M_2 P_{Y_1} i_{Y^{e,1}} & N_2 & 0 \\ M_3 P_{Y_2} i_{Y^{e,1}} & 0 & N_3 \end{pmatrix} ; \\[2em]
M^e = \begin{pmatrix} M_1 \\ M_2 P_{Y_1} i_{Y^{e,1}} \\ M_3 P_{Y_2} i_{Y^{e,1}} \end{pmatrix} ; \\[2em]
L^e = \begin{pmatrix} L_1, & P_{\mathbb{R}^m} i_{U_1} L_2, & P_{\mathbb{R}^m} i_{U_2} L_3 \end{pmatrix} \quad \text{and} \\[1em]
K^e = K_1 .
\end{cases}
$$

where $Y^{e,1} := \mathbb{R}^p \oplus W_1$.

$\square$

Remark:

In [2] Basile, Marro and Piazzi solve the DDPMS in a very simple way, however they constructed a compensator with the same dimension as the original system. This is for us undesirable since the compensator would become infinite dimensional. The compensator would also become infinite dimensional if the condition $dim(V/S) < \infty$ in *iii*) of theorem VI.15 is no longer satisfied. However for this case we could always prove a theorem similar to theorem V.11, but then we have to assume that $V_{(A,B)}(Ker\, D)$ and $V_{(A^*,C^*)}(Ker\, E^*)$ are closed subspaces, see theorem V.11.

APPENDIX E: EXAMPLES

In this appendix we shall give examples of some facts that are stated in the previous chapters. Many of the examples that will be discussed here are based on spectral realisations of delay equations. The reason for using this realisation is that on the one hand we have that these transfer functions are of a simple form and on the other hand we have from chapter IV a complete characterization of all controlled invariant subspaces. So we have the best of both worlds.

In the first section we shall recapitulate the theory of spectral realisations for delay systems as presented in Curtain & Zwart [10] and [12]. In the subsequent sections we shall give many (counter) examples of which we shall present a list.

Fact	Example
$V^*(K)$ need not necessarily exist	E.9
$V^*(K)$ may exists, and be unequal to $V_\Sigma(K)$	E.10
$V^*(K)$ may exists, and be unequal to $V_{ol}(K)$	E.11
The closed sum of two controlled invariant subspaces need not remain controlled invariant	E.15

List of examples.

Section E.1: Spectral Realisations of Delay Equations.

The delay transfer functions that will be considered are of the form

(e.1) $$f(s) = \frac{p(s, e^{-s})}{q(s, e^{-s})}$$

where $p(x, y)$ and $q(x, y)$ are polynomials in two variables. Furthermore we shall assume that the lowest power of y in $q(x, y)$ is zero. With a polynomial in two variables we define the distribution diagram.

Definition E.1: Distribution Diagram

Let $q(s,e^{-s})$ by a polynomial in two variables, then we can rewrite it in the following form

$$(e.2) \qquad q(s,e^{-s}) = \sum_{i=0}^{n} q_i(s)e^{-\beta_i s}$$

where $q_i(s)$ is a polynomial of degree m_i and $0=\beta_0<\beta_1<..<\beta_n$.

With the points (β_i,m_i) we can define the distribution diagram of $q(.,.)$; this is the polygonal line L which is the upper boundary part of the convex hull of the points $(0,0)$, (β_i,m_i), $(\beta_n,0)$.

Since this definition is rather technical we shall give a simple example to illustrate it.

Example E.2.

Consider the function $q(s,e^{-s}):=s(s+e^{-s})$. Then $(\beta_0,m_0)=(0,2)$ and $(\beta_1,m_1)=(1,1)$. So the distribution diagram has the following form.

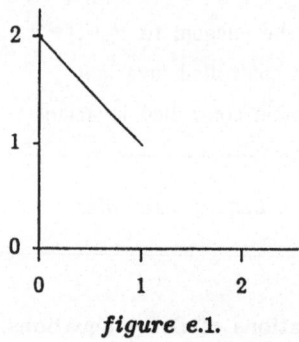

figure e.1.

□

The distribution diagram is closely related with the asymptotic distribution of the zeros of the function $q(s,e^{-s})$. To every slope of the distribution diagram there are infinitely many zeros of $q(s,e^{-s})$.

Lemma E.3.

Assume that there is no horizontal slope in the distribution diagram of the function $q(s,e^{-s})$ and assume further that on every segment of L there lie exactly two points. Let L_i be a line segment of the distribution diagram L with left end point (β_l,m_l) and right end point (β_r,m_r). So L has a kink at these points. Then the asymptotic distribution of the zeros related with this segment is given by $z_{n,j} = Re(z_{n,j}) + i\mathcal{I}m(z_{n,j})$, where

$$(e.3) \begin{cases} Re(z_{n,j}) = m\left[log(|w_j|) - log(|2\pi\, n\, m + m\, arg(w_j) \mp m\pi/2|)\right] \\ \\ \mathcal{I}m(z_{n,j}) = m\left[2\pi\, n + arg(w_j) \mp \pi/2\right]; \quad n \in \mathbb{N}, \quad 0 < j \le |m_l - m_r| \text{ and} \end{cases}$$

$m = \dfrac{m_l - m_r}{\beta_r - \beta_l}$; and w_j is j'th complex $m_l - m_r$'th root of $\dfrac{-p_{r,m_r}}{p_{l,m_l}}$; p_{i,m_i} is the leading coefficient in $p_i(s)$.

Proof:

See Bellman and Cooke [3, th. 12.10.]. $\qquad\qquad\qquad\qquad\qquad\qquad\qquad$ □

The next lemma will play a key role in obtaining spectral factorizations of delay transfer functions. Before we can state this lemma we need the concept of vertical distance.

Definition E.4: Vertical Distance.

Let L denote the distribution diagram of the function $q(s, e^{-s})$. By the vertical distance of the pair (α, d) to the distribution diagram L we shall denote the number $\tau \in \mathbb{R}$ such that $(\alpha, d + \tau) \in L$. When such a number does not exist, then the vertical distance is by definition minus infinity.

Remark:

By the definition of L we have that τ is well defined.

For example if L is the distribution diagram of $q(s, e^{-s})$, where $q(.)$ is as in example E.2, then the vertical distance of $(0,2)$ to L is 0 and the vertical distance of $(2,0)$ is $-\infty$.

With the concept of vertical distance we can state the main result of Curtain and Zwart [10].

Lemma E.5.

Let $q(s,e^{-s})$ be the same as in (e.2) and let L denote the distribution diagram of q. Suppose that on each segment of L lie exactly two points (β_i, m_i).

Let $p(s,e^{-s}) = \sum_{i=1}^{n_p} p_i(s)e^{-\alpha_i s}$ with $p_i(s)$ a polynomial of degree d_i, and assume that p and q are coprime. If the vertical distance of the points (α_i, d_i) to L is larger than zero for $1 \leq i \leq n_p$, then $f(s) = \dfrac{p(s,e^{-s})}{q(s,e^{-s})}$ has a partial fraction expansion, given by

$$(e.4) \qquad f(s) = \sum_{j=m+1}^{\infty} \frac{p(z_j)/q'(z_j)}{(s-z_j)} + \sum_{j=0}^{m} \sum_{i=0}^{n_j} \frac{c_{i,j}}{(s-z_j)^i},$$

where z_j are the zeros of $q(s)$ with multiplicity n_j, and this sum is uniformly convergent in s in any compact subset of $\mathbb{C}/\{z_j\}$.

Furthermore we have that $|p(z_j)/q'(z_j)| \leq C|z_j|^{-\tau}$, where C is independent of z_j and τ is the minimal vertical distance of the points (α_i, d_i) to L.

Proof:

See Curtain and Zwart [10, p. 73 and 74]. $\qquad\qquad\qquad\qquad\qquad\qquad$ □

This theorem enables us to make spectral realisations of delay transfer functions. In order to improve the readability we shall list the required properties of $f(s) = \dfrac{p(s,e^{-s})}{q(s,e^{-s})}$.

Definition E.6: Property P.

Let $q(s,e^{-s}) = \sum_{i=0}^{n} q_i(s)e^{-\beta_i s}$, where $q_i(s)$ is a polynomial of degree m_i and $0 = \beta_0 < \beta_1 < .. < \beta_n$ and let L denote the distribution diagram of q.

Let $p(s,e^{-s}) = \sum_{i=1}^{n_p} p_i(s)e^{-\alpha_i s}$ with $p_i(s)$ a polynomial of degree d_i, and assume that p and q are coprime i.e. no common zeros. Then $f(s) = \dfrac{p(s,e^{-s})}{q(s,e^{-s})}$ satisfies property P if it satisfies $P1$, $P2$ and $P3$, where

P1.   On each segment of L lie exactly two points (β_i, m_i).

P2.   If (β_l, m_l) and (β_r, m_r) are two succeeding points on L with $\beta_l < \beta_r$, then $m_l - m_r = 1$.

P3.   Let τ_i denote the vertical distance of the point (α_i, d_i) to L. Then the minimum of these τ_i's is larger than one.

Note that $P2$ implies that there are no horizontal segments in L, and furthermore it implies that to every segment of L there is one string of zeros, see $(e.3)$.

With the property P we can make a spectral realisation. In order to shorten the notation we shall assume that q has no multiple zeros. We stress that this assumption is made only to improve the readability. The same results as stated in the next theorem holds if this condition would be omitted.

Theorem E.7.

If $f(s) = \dfrac{p(s, e^{-s})}{q(s, e^{-s})}$ has property P and assume that all the poles of f have multiplicity one, (see the remark above), then there exists a spectral realisation (D, A, B) with state space ℓ^2 such that $f(s) = D(s-A)^{-1}B$, where

$$Ax = \sum_{n=1}^{\infty} z_n < x, e_n > e_n;$$

$$D(A) = \{x \in \ell^2 | \sum_n |z_n|^2 | < x, e_n > |^2 < \infty\}; \ z_n \ \text{are the zeros of } q.$$

$$Bu = bu \ \text{and} \ Dx = <x, d>; \ b = d = \{(p(z_n)/q'(z_n))^{1/2}\}$$

Furthermore this realisation has the following properties from chapter IV:

(Δ1)   The generator A is a discrete spectral operator.

(Δ2) $b_i := \ <b, e_i> \ \neq 0$, for all $i \geq 1$.

(Δ3)   For all $i \in \mathbb{N}$, $De_i \neq 0$.

(Δ4) $\inf\limits_{i \neq j} |z_i - z_j| = \delta > 0$,

(Δ5) $\sup\limits_{i \in \mathbb{N}} \sum\limits_{\substack{j=1 \\ i \neq j}}^{\infty} \left| \dfrac{1}{z_i - z_j} \right|^2 < \infty.$

We remark that $\sigma(A) = \{z_j\}$.

Proof:

From property $P3$ and lemma E.5 we have that $|p(z_n)/q'(z_n)| < Cn^{-\tau}$ with $\tau > 1$. So b and d are elements of ℓ^2. That (D, A, B) is a spectral realisation of $f(s)$ follows now easily from lemma E.5. Now we shall prove that this realisation satisfies $\Delta1$ up to $\Delta5$.

The resolvent of A is given by

$(e.5)$ $$(\lambda - A)^{-1} = \sum_{n \in \mathbb{N}} \dfrac{1}{\lambda - z_n} <., e_n> e_n$$

and from $(e.3)$ we have that $|z_n| < c(n)$. Thus the resolvent is compact, and so we have proved $\Delta 1$. $\Delta 2$ and $\Delta 3$ follow easily from the definition of b and d and the condition that p and q are coprime.

Furthermore we have from $(e.3)$ that condition $\Delta 4$ holds and we have that fixed $i \in N$ $|z_i - z_j| > c^* j$; $i \neq j$ with $c > 0$ independent of i. So $\Delta 5$ is also satisfied

$\square$

Remark:

The properties $\Delta 1$, $\Delta 4$ and $\Delta 5$ are properties of the system matrix. So if b and d is another pair such that $f(s) = \,<(s-A)^{-1}b, d>$ and b satisfies $\Delta 2$ and d satisfies $\Delta 3$, then this realisation satisfies also $\Delta 1$ up to $\Delta 5$.

We shall illustrate this theorem by the following easy example.

Example E.8.

Consider the transfer function $f(s) = \dfrac{1}{4s(-4s + e^{-4s})}$. Then by theorem E.7 we have that the spectral realisation of f is given by

$$Ax = \sum_{n=1}^{\infty} z_n <x, e_n> e_n;$$

$$D(A) = \{x \in \ell^2 | \sum_n |z_n|^2 |<x, e_n>|^2 < \infty\}; \quad z_n, \ n > 1 \text{ are the zeros of } (-4s + e^{-4s});$$

$$z_1 = 0 \text{ and}$$

$$b = d = \{1/2, 1/(16z_2(-1-4z_2))^{1/2}, \ldots, 1/(16z_j(-1-4z_j))^{1/2}, \ldots\}.$$

Furthermore this realisation satisfies $\Delta 1$ up to $\Delta 5$, and so we have from theorem IV.16 a complete characterization of all controlled invariant subspaces in the kernel of D. Since the transfer function has no zeros we have from theorem IV.16 that the unique controlled invariant subspace in the kernel of D is the zero subspace. So we have for this realisation that $V^*(Ker\, D)$ exists and it is equal to the zero subspace. Note that we do not state anything about the more usual realisation in the space M^2.

$\square$

In the next section we shall use these spectral realisations of delay equations in order to obtain counter examples for properties that hold if the state space is finite dimensional but do not alway hold for the infinite dimensional case.

Section E.2: The Relation between $V^*(K)$, $V_{ol}(K)$ and $V_\Sigma(K)$

The next example will show that $V^*(K)$ need not exist.

Example E.9.

Let $\mathcal{H}$ be $L^2(0,1)$ and A is the "heat operator", $A = \dfrac{d^2}{dz^2}$ with domain,

$D(A) = \{x \in L^2(0,1) | x'' \in L^2(0,1); \ x(0) = x(1) = 0\}$,

$$B := b(z) = \begin{cases} 0 & ; \ z \in [0, \tfrac{1}{2}] \\ \\ sin(2\pi z); & z \in [\tfrac{1}{2}, 1] \end{cases}$$

$K = \{x \in L^2(0,1) | \ x(z) = 0; \ a.e. \ on \ [0, \tfrac{1}{2}]\}$.

Notice that $K \supset Im \ B$. So $B^0(K) = \{0\}$, and from lemma II.19 we may conclude that, if K is controlled (or equivalently frequency) invariant, then every x in K has an unique (ξ, ω) representation with $B\omega(s) \equiv 0$, which means that $(s-A)^{-1}K \subset K$. Thus K would be $T_A(t)$-invariant, but this is in contradiction with example I.6. So K cannot be controlled invariant.

Define for $n > 1$

$$e_n(z) = \begin{cases} 0 & z \in [0, \tfrac{1}{2}] \\ \\ \dfrac{1}{4\pi^2 n(-1)^n(n^2 - 1)} \{sin(2\pi n \ z) + n(-1)^n sin(2\pi \ z)\}; & z \in [\tfrac{1}{2}, 1] \end{cases},$$

and μ_n in $\mathbb{C}$ by $\mu_n = -4\pi^2 n^2$.

Then   1)   span$\{e_i\}_{i=2..n}$ is controlled invariant for all $n \in \mathbb{N}$

       2) $\overline{span}\{e_i\}_{i \in \mathbb{N}/\{1\}}$ is K.

Proof of 1):

By a simple calculation one can show that $(A - \mu_n)e_n = b$. So

(e.6) $(s-A)e_n = -b + (s - \mu_n)e_n$ or $e_n = (s-A)\dfrac{e_n}{s - \mu_n} - b\dfrac{1}{\mu_n - s}$.

Thus span$\{e_n\}$ is controlled invariant. By lemma III.14 we have that span$\{e_i\}_{i=2..n}$ is controlled invariant for all $n \in \mathbb{N}$.

Proof of 2):

If $x \in K$, then $x = \begin{cases} 0 \ on \ [0, \tfrac{1}{2}] \\ \\ \tilde{x} \ on \ [\tfrac{1}{2}, 1] \end{cases}$, with $\tilde{x} \in L^2(\tfrac{1}{2}, 1)$. So

$$\tilde{x}(z) = \sum_{n=1}^{\infty} <\tilde{x}(z), 2sin(2\pi n \ z)>_{L^2(\frac{1}{2},1)} 2sin(2\pi n \ z) \ \ and$$

$$\|\tilde{x}\|^2 = \sum_{n=1}^{\infty} |<\tilde{x}(z), 2sin(2\pi n \ z)>_{L^2(\frac{1}{2},1)}|^2 < \infty$$

Let $x \perp e_n \ \forall \ n > 1 \Rightarrow <x, e_n>_{L^2(\frac{1}{2},1)} = 0 \Rightarrow$

$<\tilde{x}(z), sin(2\pi n \ z) + n(-1)^n sin(2\pi \ z)>_{L^2(\frac{1}{2},1)} = 0.$ So

(e.7) $<\tilde{x}, sin(2\pi n \ z)>_{L^2(\frac{1}{2},1)} = -n(-1)^n <\tilde{x}, sin(2\pi \ z)>_{L^2(\frac{1}{2},1)} \ \forall \ n > 1$

(e.8) $\infty > \|\tilde{x}\|^2 \geq \sum_{n=2}^{\infty} |<\tilde{x}, sin(2\pi n \ z)>_{L^2(\frac{1}{2},1)}|^2 = \sum_{n=2}^{\infty} n^2 |<\tilde{x}, sin(2\pi \ z)>_{L^2(\frac{1}{2},1)}|^2$

(e.8) implies that $<\tilde{x}, sin(2\pi \ z)>_{L^2(\frac{1}{2},1)} = 0$, and with (e.7) we have that $<\tilde{x}, sin(2\pi n \ z)>_{L^2(\frac{1}{2},1)} = 0$, for all n in $\mathbb{N}/\{0\}$, so $\tilde{x} = 0$.

Thus $x \equiv 0$ is the only vector in K perpendicular on all e_n, so $\overline{span}_{i \in \mathbb{N}/\{1\}} \{e_i\}$ is K.

If $V^*(K)$ were to exist, then it would necessarily be closed, contained in K and by 1) it must contain $span_{i=2..n} \{e_i\}$, for all $n \in \mathbb{N}$. This together with 2) would imply that $V^*(K)$ equals K. This contradicts the fact that K is not controlled invariant. Thus $V^*(K)$ cannot exist in this example.

□

The next example will show that it is possible that $V^*(K)$ exists but it is unequal to $V_\Sigma(K)$. Note that theorem III.12 implies that $V_\Sigma(K)$ cannot be closed then.

Example E.10.

In this example we shall study the delay transfer function as introduced in example E.8. So

(e.9) $$f(s) = \frac{1}{4s(-4s + e^{-4s})}$$

From example E.8 we have that this system has a spectral realisation (D, A, B) which satisfies the conditions $\Delta 1$ up to $\Delta 5$ and we have that D is given by $D = <.,d>$ where $d = \{1/2, 1/\left(16z_2(-1-4z_2)\right)^{1/2}, .., 1/\left(16z_j(-1-4z_j)\right)^{1/2}, .\}$; z_j, $j \geq 2$ is the $j-1$ th zero of $-4s + e^{-4s}$. Furthermore we have proved that $V^*(Ker\, D)$ is the zero subspace and thus by theorem III.3 the DDP is only solvable if $E = 0$, i.e. no disturbances.

Now we shall show that there exists a bounded operator E and a strictly proper $U(s) \in \left[\mathcal{L}(Q, \mathbb{C})\right]_{-1}(s)$ such that

$(e.10)$ $\qquad D(s-A)^{-1}BU(s) = D(s-A)^{-1}E.$

The existence of such a disturbance input operator will prove two facts. Firstly; since the D.D.P. is non solvable, but the meromorphic matrix equations (3.13), (3.14), and (3.15) are solvable, we have that the solvability of these equations is in general not equivalent to the solvability of DDP. So the condition that $V_{\Sigma}(Ker\, D)$ is closed can not be omitted in theorem III.10. It easy to prove that the solvability of (3.13), (3.14) or (3.15) is always a necessary condition for the solvability of DDP.

Secondly, we have that $V^*(Ker\, D)$ can exists, and it is unequal to $V_{\Sigma}(Ker\, D)$. Namely from equation $(e.10)$ we have that Eq is contained in $V_{\Sigma}(Ker\, D)$, since

$(e.11)$ $\qquad D(s-A)^{-1}Eq = D(s-A)^{-1}BU(s)q,$ so

$\qquad (s-A)^{-1}Eq = \xi(s) + (s-A)^{-1}BU(s)q,$ with $\xi(s) \in Ker\, D.$

Thus

$(e.12)$ $\qquad Eq = (s-A)\xi(s) + BU(s)q$

and by definition this shows that $Eq \in V_{\Sigma}(Ker\, D)$, and thus $\{0\} \neq V_{\Sigma}(Ker\, D)$.
Now we shall define this operator E.

Let $E: \mathbb{C} \rightarrow \ell^2$ be defined by $Eq = eq$, where

$(e.13)$ $\qquad e = \{1/2, 1/\left[2(4z_2)^{1/4}(-1-4z_2)^{1/2}\right], .., 1/\left[2(z_j)^{1/4}(-1-4z_j)^{1/2}\right], ..\}$

From $(e.3)$ we have that the j th component of e is of the order $j^{-\frac{3}{4}}$, so $e \in \ell^2$, and thus E is a bounded operator.

Furthermore it is from lemma E.5 easy to see that (D, A, E) is a realisation of the transfer function

$$(e.14) \qquad D(s-A)^{-1}E = \frac{e^{-s}}{4s(-4s+e^{-4s})}$$

If we define $U(s)$ as e^{-s}, then the operator E and the function $U(s)$ satisfies equation $(e.10)$ and so we have constructed the counter example. $\qquad \square$

Similar to this example we shall construct an example such that $V^*(Ker\,D)$ exists, but it is unequal to the largest open loop invariant subspace in the kernel of D, $V_{ol}(Ker\,D)$.

Example E.11.

Again we shall consider the spectral realisation of the delay transfer function $f(s) = \dfrac{1}{4s(-4s+e^{-4s})}$. By example E.8 we have that the spectral realisation of f is given by

$$Ax = \sum_{n=1}^{\infty} z_n <x,e_n> e_n;$$

$D(A) = \{x \in \ell^2 | \sum_n |z_n|^2 | <x,e_n>|^2 < \infty\}$; z_n, $n>1$ are the zeros of $(-4s+e^{-4s})$;

$z_1 = 0$ and

$$b = d = \{1/2, 1/\left(16z_2(-1-4z_2)\right)^{1/2}, .., 1/\left(16z_j(-1-4z_j)\right)^{1/2}, ..\}.$$

Now we shall construct an initial value $x_0 \in Ker\,D$ such that there exists a continuous input function $u(t)$ such that the solution of $\dot{x}(t) = Ax(t) + bu(t)$; $x(0) = x_0$ remains in the kernel of D. As x_0 we take

$$(e.14) \qquad x_0 = \left\{(1+2a)/2, \left(-(4z_2)^{1/4} - 2a(z_2)^{1/2} - a)\right) / \left((z_2^2-1)(16z_2(-1-4z_2))^{1/2}\right), . \right.$$

$$\left. ., \left(-(4z_j)^{1/4} - 2a(z_j)^{1/2} - a)\right) / \left((z_j^2-1)(16z_j(-1-4z_j))^{1/2}\right), ... \right\}$$

where

$$(e.15) \qquad a = \frac{sinh(-1)}{sinh(2)}$$

and with $u(t)$ defined as

$$(e.16) \quad u(t)= \begin{cases} a\ sinh(t) & ;0\le t\le 1 \\ sinh(t-1)+a\ sinh(t) & ;1\le t\le 2 \\ 0 & ;t\ge 2 \end{cases}$$

we claim that this input will keep x_0 in the kernel of D.

So we have to prove that

$$(e.17) \qquad 0=Dx(t)=DT_A(t)x_0+D\int_0^t T_A(t-s)Bu(s)ds; \ \forall t\ge 0.$$

First we remark that, by the definition of a, $u(.)$ is a continuous function on $[0,\infty)$. Furthermore the Laplace transform of $u(.)$ is given by

$$(e.18) \qquad \mathcal{L}(u)(s): =\omega(s)=\frac{e^{-s}+a\,e^{-2s}+a}{s^2-1}$$

Taking the Laplace transform of (e.17) yields

$$(e.19) \qquad 0=D(s-A)^{-1}x_0+D(s-A)^{-1}B\omega(s)$$

So from our realisation it thus remains to show that

$$(e.20) \qquad D(s-A)^{-1}x_0= -\frac{1}{4s(-4s+e^{-4s})}\ \frac{e^{-s}+a\,e^{-2s}+a}{s^2-1}.$$

Or, equivalently, we must prove that (D,A,x_0) is a realisation of

$$(e.21) \qquad g_a(s)= -\frac{e^{-s}+a\,e^{-2s}+a}{4s(s^2-1)\,(-4s+e^{-4s})},$$

This result follows directly from the next lemma. Thus $x_0\in V_{ol}(Ker\ D)$, and so we have constructed our example. $\qquad\qquad\qquad\qquad\qquad\qquad\qquad \square$

Lemma E.12.

Consider the function $g_a(s)=\dfrac{-e^{-s}-a\,e^{-2s}-a}{4s(s^2-1)(-4s+e^{-4s})}$, with $a=\dfrac{sinh(-1)}{sinh(2)}$. Then on $\mathbb{C}$ the following equality holds: $g_a(s)=D(s-A)^{-1}x_0$, where D, A are as defined in example E.8 and x_0 is defined by (e.14).

Proof:

Since the numerator and the denominator have by the special value of a, common zeros we can not apply theorem E.7 directly. However if we define the sequence $\{a_n\}\subset\mathbb{R}$ such that $a_n\to a$ and $-e^{-s}-a_ne^{-2s}-a_n$ has no common zeros with the denominator of $g_a(.)$. Then we can construct a spectral realisation of

(e.22)
$$g_{a_n}(s) := \frac{-e^{-s} - a_n e^{-2s} - a_n}{4s(s^2-1)(-4s+e^{-4s})}$$

As state space for these realisation we choose $\ell^2(\{-1,0,1..\})$. This space is obviously isometric isomorph with $\ell^2(\mathbb{N})$, but we shall need this space for notational convenience.

From theorem E.7 we have that a realisation of $g_{a_n}(s)$ is given by

$$A^e x = \sum_{n=-1}^{\infty} z_n <x, e_n> e_n;$$

$$D(A) = \{x \in \ell^2(\{-1,0,..\}) | \sum_{-1}^{\infty} |z_n|^2 | <x, e_n> |^2 < \infty\};$$

$z_{-1} = -1, z_0 = 1,$ $z_1 = 0$ and $z_n,$ $n \geq 2$ are the zeros of $(-4s + e^{-4s})$;

$$d^e = \{1, 1, 1/2, 1/(16z_2(-1-4z_2))^{1/2}, .., 1/(16z_j(-1-4z_j))^{1/2}, ..\}.$$

$$x^n = \left\{ \frac{(e+a_n e^2 + a_n)}{8(4+e^4)}, \frac{(e^{-1} + a_n e^{-2} + a_n)}{8(4-e^{-4})}, \frac{(1+2a_n)}{2}, \frac{-(4z_2)^{1/4} - 2a_n(z_2)^{1/2} - a_n}{4(z_2^2-1)(z_2(-1-4z_2))^{1/2}}, .. \right.$$

$$\left. .., \frac{-(4z_j)^{1/4} - 2a_n(z_j)^{1/2} - a_n}{(z_j^2-1)(16z_j(-1-4z_j))^{1/2}}, .. \right\}$$

If n converges to infinity, then $g_{a_n}(s)$ converges for fixed $s \in \mathbb{C}$ to $-g_a(s)$. Furthermore it is easy to see that x^n converges to $x_0^e := \{0, 0, x_0\}$ in the norm of $\ell^2(\{-1,0,1,..\})$. So for fixed $s \in \mathbb{C}$ $D^e(s-A^e)^{-1}x^n$ converges to $D^e(s-A^e)^{-1}x_0^e = D(s-A)^{-1}x_0$. So for fixed $s \in \mathbb{C}$ we have that $g_a(s) = D(s-A)^{-1}x_0$. $\square$

Section E.3: On the Sum of two Controlled Invariant Subspaces

In this section we shall show by means of a counter example that the closure of the sum of two controlled invariant subspaces is not necessarily controlled invariant. In this example we shall use the spectral realisation of a delay equation as derived in section E.1. The delay equation that will be considered is given by

(e.23)
$$F(s) = \frac{(se^{-s} + s^2 + 1)(se^{-s} + s^2 - 1)}{(s^4-1)(e^{-s}+s)(e^{-2s}+s)}$$

We shall prove that the spectral realisation of this equation satisfies the conditions of chapter 4. Furthermore we shall prove that the subspace associated with the zeros of $se^{-s} + s^2 + 1$ as well as the subspace associated with the zeros of $se^{-s} + s^2 - 1$ is controlled invariant, (see theorem IV.16). However the sum of these subspaces will not remain controlled invariant, since to every pole of $F(s)$ associated with $(e^{-s} + s)$ there are two zeros of the system, see theorem IV.16.

Before we can prove our example we need some results on the distribution of the poles and zeros of $F(s)$.

Lemma E.13.

Let $h(.)$ and $\eta(.)$ satisfies on $\mathbb{C}$ the following relation

$$(e.24) \qquad h(s) = h(s_0) + (s - s_0)h'(s_0) + (s - s_0)\eta(s).$$

where $h'(s_0) \neq 0$.

Assume further that on the circle $C := \{z: \ |z - s_0| = 2 \dfrac{|h(s_0)|}{|h'(s_0)|}\}$ the following inequality holds $|\eta(z)| < \tfrac{1}{2}|h'(s_0)|$. Then $h(.)$ has exactly one zero inside C.

Proof:

Define $f(s) := h(s) - (s - s_0)h'(s_0)$ and $g(s) := (s - s_0)h'(s_0)$. Then for $z \in C$ we have that

$$|g(z)| = |z - s_0||h'(s_0)| = \tfrac{1}{2}|z - s_0||h'(s_0)| + \tfrac{1}{2}|z - s_0||h'(s_0)| =$$

$$= |h(s_0)| + \tfrac{1}{2}|z - s_0||h'(s_0)| > |h(s_0)| + |z - s_0||\eta(z)| >$$

$$|h(s_0) + (z - s_0)\eta(z)| = |f(z)|, \text{ by } (e.24) \text{ and the definition of } f(.).$$

Now with Rouché theorem, Rudin [31] we have that $g(.)$ and $f + g = h$ have the same number of zeros inside C. So $h(.)$ has exactly one zero inside C. $\qquad \square$

From this lemma we obtain the implication concerning the distance between the zeros and poles of $F(s)$.

Corollary E.14.

Let s_0 be a zero of $e^{-s} + s$ with norm sufficiently large. Then inside a circle with centre s_0 and radius $\dfrac{2}{|s_0(s_0 + 1)|}$ there is exactly one zero of $se^{-s} + s^2 + 1$.

Proof:

Defining $h(s): = se^{-s}+s^2+1$ and $\eta(s): = \dfrac{se^{-s}+s^2-(s-s_0)s_0(s_0+1)}{(s-s_0)}$ gives that this pair satisfies $(e.6)$. Furthermore we have from $(e.6)$ and the fact that h and η are entire function that

$(e.25)$ $$\eta(s) = \sum_{k=2}^{\infty} \frac{h^{(k)}(s_0)}{k!}(s-s_0)^{k-1}; \quad s \in \mathbb{C}.$$

Calculating $h^{(n)}(s_0)$ gives; $h(s_0)=1$; $h'(s_0)=s_0(s_0+1)$; $h''(s_0) = -s_0^2+2s_0+2$ and $h^{(n)}(s_0)=n(-1)^n s_0+(-1)^{n+1}s_0^2$;$n>2$. So

$(e.26)$ $$\eta(s) = \sum_{k=2}^{\infty} \frac{(-1)^{k+1}s_0^2}{k!}(s-s_0)^{k-1} + \sum_{k=2}^{\infty} \frac{k(-1)^k s_0}{k!}(s-s_0)^{k-1}+(s-s_0) =$$

$$s_0^2 \sum_{k=1}^{\infty} \frac{(-1)^k}{(k+1)!}(s-s_0)^k + s_0 \sum_{k=1}^{\infty} \frac{(-1)^{k+1}}{k!}(s-s_0)^k+(s-s_0).$$

With equation $(e.26)$ we have that

$(e.27)$ $$|\eta(s)| \le |s_0|^2 \sum_{k=1}^{\infty} \frac{|s-s_0|^k}{(k+1)!} + |s_0| \sum_{k=1}^{\infty} \frac{|s-s_0|^k}{k!} +|s-s_0| =$$

$$= |s-s_0||s_0|^2 \sum_{k=0}^{\infty} \frac{|s-s_0|^k}{(k+2)!} +|s-s_0||s_0| \sum_{k=1}^{\infty} \frac{|s-s_0|^k}{(k+1)!} +|s-s_0| <$$

$$< |s-s_0|\left\{ \tfrac{1}{2}|s_0|^2 e^{|s-s_0|}+|s_0|e^{|s-s_0|}+1 \right\}.$$

So on the circle $|z-s_0| =2\dfrac{|h(s_0)|}{|h'(s_0)|}= \dfrac{2}{|s_0(s_0+1)|}$ we have that

$(e.28)$ $$|\eta(z)| < \frac{1}{|s_0(s_0+1)|}\left\{ |s_0|^2 e^{\frac{2}{|s_0(s_0+1)|}} +2|s_0|e^{\frac{2}{|s_0(s_0+1)|}} +2 \right\}.$$

and for $|s_0|$ sufficiently large this is smaller than $\tfrac{1}{2}|s_0(s_0+1)| = \tfrac{1}{2}|h'(s_0)|$. So lemma E.13 gives the desired result. $\square$

Remark:

With a similar proof one can show that the zeros of the function $se^{-s}+s^2-1$ have the same property.

Now we have proved the necessary lemmas to give a counter example for the fact that the closure of the sum of two controlled invariant subspace is not necessarily controlled invariant.

Example E.15.

Consider the delay transfer function

$$F(s) = \frac{(se^{-s} + s^2 + 1)(se^{-s} + s^2 - 1)}{(s^4 - 1)(e^{-s} + s)(e^{-2s} + s)}$$

We shall begin by constructing a spectral realisation of this transfer function. Let $\{z_j; j > 0\}$ denote the zeros of $e^{-s} + s$, and $\{z_j; j < -3\}$ denote the zeros of $e^{-2s} + s$, and $z_0 = 1$, $z_{-1} = -1$, $z_{-2} = i$, $z_{-3} = -i$. So $\{z_j; j \in \mathbf{Z}\}$ are the poles of $F(s)$.

The spectral realisation of $F(s)$ will be constructed on $\ell^2(\mathbf{Z})$. Since this space is isometrically isomorph with $\ell^2(\mathbf{N})$ we have that a realisation on $\ell^2(\mathbf{Z})$ gives by renumbering of the indices also a realisation on $\ell^2(\mathbf{N})$. So the realisation on $\ell^2(\mathbf{Z})$ has similar properties as the realisation on $\ell^2(\mathbf{N})$ i.e. if the transfer function satisfies the conditions in theorem E.7., then the realisation on $\ell^2(\mathbf{Z})$ will satisfy $\Delta 1, .., \Delta 5$.

On $\ell^2(\mathbf{Z})$ we have the following realisation of $F(s)$;

$$Ax = \sum_{n \in \mathbf{Z}} z_n <x, e_n> e_n;$$

$$D(A) = \{x \in \ell^2(\mathbf{Z}) | \sum_{n \in \mathbf{Z}} |z_n|^2 | <x, e_n> |^2 < \infty\};$$

$$z_n, \ j < -3 \text{ are the zeros of } e^{-2s} + s;$$

$$z_{-3} = -i, \ z_{-2} = i, \ z_{-1} = -1, \ z_0 = 1, \text{ and}$$

$$z_n, \ n \geq 1 \text{ are the zeros of } (s + e^{-s});$$

$$b = \{b_j\}_{j \in \mathbf{Z}}; \qquad b_j = \frac{1}{z_j + 1}; \ j \in \mathbf{N},$$

$$b_j = 1; \ j = -3, .., 0$$

$$b_j = \frac{1}{2z_j + 1}; \ j < -3.$$

$$D = <.,d>; \quad d = \{d_j\}; \quad d_j = \frac{-1}{(z_j^4 - 1)(z_j^2 + z_j)} \quad j \geq 1;$$

$$d_0 = \frac{(e^{-1} + 2)(e^{-1})}{4(e^{-1} + 1)(e^{-2} + 1)}, \quad d_{-1} = \frac{(-e^1 + 2)(-e^1)}{-4(e^1 - 1)(e^2 - 1)},$$

$$d_{-2} = \frac{(ie^{-i})(ie^{-i} - 2)}{4i(e^{-i} + i)(e^{-2i} + i)}, \quad d_{-3} = \frac{(-ie^i)(-ie^i - 2)}{-4i(e^i - i)(e^{2i} - i)},$$

$$d_j = \frac{z_j^2 \left((-z_j)^{1/2} + z_j \right)^2 - 1}{(z_j^4 - 1) \left((-z_j)^{1/2} + z_j \right)}, \quad j < -3.$$

From theorem E.7 we have that (D, A, b) is a spectral realisation of $F(.)$ and furthermore it has the properties $\Delta 1, .., \Delta 5$ from chapter IV. So we have in this realisation a complete characterization of all controlled invariant subspaces.

Now we shall construct two subspaces which are controlled invariant, but whose closed sum is no longer controlled invariant. From chapter IV we have that controlled invariant subspaces are closely related to the zeros of the transfer function. Corollary E.14 gives that outside a sufficiently large circle Γ there is with every zero of $(e^{-s} + s)$ exactly one zero of $se^{-s} + s^2 + 1$ and also one zero of $se^{-s} + s^2 - 1$. Let $\{\mu_j\}_{j \in \mathbb{N}}$ and $\{v_j\}_{j \in \mathbb{N}}$ denote these zeros of respectively $se^{-s} + s^2 + 1$ and $se^{-s} + s^2 - 1$. With these zeros we define the following closed subspaces

$$(e.29) \qquad V_1 = \overline{span}\{(\mu_j - A)^{-1}b\}_{j \in \mathbb{N}}$$

$$(e.30) \qquad V_2 = \overline{span}\{(v_j - A)^{-1}b\}_{j \in \mathbb{N}}$$

Now we shall show that V_1 and V_2 are controlled invariant. This we shall only do for V_1, the proof for V_2 is similar.

By definition IV.12 we have that V_1 is of the form (4.9). So V_1 satisfies condition $a)$ of theorem IV.16, furthermore it also satisfies condition $b)$. Thus it remains to prove the existence of a subsequence n_j such that $\sum_{j \in \mathbb{N}} \left[\frac{z_{n_j} - \mu_j}{b_{n_j}} \right]^2 < \infty$. As the sequence of poles we take those poles that are closest to the zeros. Then we have by definition of μ_j that this pole is z_{j+k}, k is the number of zeros of $e^{-s} + s$ inside the circle Γ and from corollary E.14 we have that $|z_{n_j} - \mu_j| = |z_{j+k} - \mu_j| < \frac{2}{|z_{j+k}(z_{j+k} + 1)|}$.

$$\text{So} \sum_{j \in \mathbb{N}} \left| \frac{z_{n_j} - \mu_j}{b_{n_j}} \right|^2 = \sum_{j \in \mathbb{N}} \left| \frac{z_{j+k} - \mu_j}{b_{j+k}} \right|^2 \leq \sum_{j \in \mathbb{N}} \left| \frac{\frac{2}{|z_{j+k}(z_{j+k}+1)|}}{\frac{1}{z_{j+k}+1}} \right|^2 =$$

$$\sum_{j \in \mathbb{N}} \left| \frac{2}{z_{j+k}} \right|^2 < \infty, \text{ by } (e.3).$$

Thus V_1 is controlled invariant.

Now we shall prove that $V_1 + V_2$ is not controlled invariant.

As a consequence of theorem IV.16, $\Delta 4$ and the fact that b_{n_j} converges to zero if j goes to infinity, we have that outside a sufficiently large circle to every pole of $F(s)$ there is one or none zero of the controlled invariant subspace. So if $V: = V_1 + V_2$ were to be controlled invariant, then there would exists to the poles z_j; $j \in \mathbb{N}$, for j sufficiently large, one or no zero. However we have that to every such pole there exists a zero of $se^{-s} + s^2 + 1$ and one of $se^{-s} + s^2 - 1$. Thus V cannot be controlled invariant.

□

CONCLUSIONS

In this monograph we have presented a fundamental treatment of the geometric theory for infinite dimensional systems in Banach spaces. We have proved that for closed subspaces the concepts of open loop invariance, closed loop invariance and frequency invariance are equivalent. The condition that the subspace must be closed is really essential; there are examples of open loop invariant subspaces which are not closed loop invariant, and the same holds for frequency invariance. There is another fundamental difference between the concepts of frequency invariance, open loop invariance and the concept of closed loop invariance. Namely the largest frequency invariant subspace contained in a given closed subspace K, $V_\Sigma(K)$, and the largest open loop invariant subspace contained in K, $V_{ol}(K)$ both exist, but in general there may not exist a largest closed loop invariant subspace contained in K, $V^*(K)$. There can concur even stranger things. It has been shown by means of an example that it is possible that both the largest closed loop invariant and the largest frequency invariant subspace exist, but are unequal. This is very unsatisfactory, since an easy test for the solvability of the DDP in terms of the transfer functions from the input to the measurement and from the disturbance to the measurement, which was valid in the finite dimensional case, loses it validity for infinite dimensional systems. This test is essentially based on the concept of (ξ,ω)–representation. This concept, although introduced by Hautus [19] for finite dimensional systems, is of great use for the infinite dimensional case. For example with this concept we proved the equivalence between open loop and closed loop invariance, but it can also be used to obtain strong results in the theory of stabilizability, Zwart [49].

As we have seen, if the largest frequency invariant subspace, $V_\Sigma(K)$, equals the largest closed loop invariant subspace, $V^*(K)$, then the theorems form a complete analogy to the finite dimensional case. So the equality between $V_\Sigma(K)$ and $V^*(K)$ is essential. For applications it is however very hard to check whether $V_\Sigma(K)$ equals $V^*(K)$. In chapter four we have focussed on the existence of $V^*(K)$ for the class of discrete spectral systems on a Hilbert space. There we proved that the existence of this subspace is equivalent to a pole placement problem, where the new poles must be placed on the zeros of the system. However since the existence of $V^*(K)$ does not imply that this subspace is equal to $\mathcal{V}_\Sigma(K)$, it is again hard to show that the DDP is solvable if this pole placement problem is solvable. From all these results we see that there is a gap between the nice mathematical theory and the applications. One could think of several ways to investigate the solvability of the DDP in practical applications.

One way of attacking this problem would be to approximate the given system by a finite dimensional one, for which one solves the disturbance decoupling problem, and then to use this solution as an approximate solution for the general system. By the non–robustness of the disturbance decoupling problem it is not clear that this approach should work. The non–robustness of the DDP can easily be seen from the following remark. For every positive number ε there exists a D_ε such that $\|D - D_\varepsilon\| < \varepsilon$ and $D_\varepsilon E \neq 0$. So by making only a small change in norm one can violate the necessary condition for the solvability of the DDP.

A better approach to the general solvability of the DDP would be to investigate its almost version as in Trentelman [39] in the hope that this solution will be more robust. In the infinite dimensional case we also covered the special case that the generator A is bounded. For continuous time systems the condition that A is bounded is too strong to include interesting examples. However if one considers discrete time systems, then most generators will be bounded. It can easily be deduced from the results in this monograph that for this case the geometric theory is completely analogous to the finite dimensional case. One can state this as "the geometric theory is a theory for bounded operators". The "almost" version of the DDP for continuous time systems is still completely open, however.

So we still have many open problems if we want to solve the DDP for applications. However since we are working in an infinite dimensional state space the assumption in the DDP that the whole state is measured is not very reasonable. In most applications the output will be a part of the state, and this brings us to the problem of DDPM or DDPMS if we also want to guarrantee internal stability. For a system defined in a Hilbert space we have derived necessary and sufficient conditions in terms of system–invariance subspaces for the solvability of the DDPM and the DDPMS with a finite dimensional compensator In these conditions we did not needed the condition that $\mathcal{V}_\Sigma(K)$ equals $\mathcal{V}^*(K)$, but this only holds for the case that the to–be–constructed compensator is finite dimensional. Furthermore these results are (still) not applicable in a general Banach space.

Concluding we can see that at this moment it is not clear what one can do about general disturbance decoupling problems for systems for which the equality between $V_\Sigma(K)$ and $V^*(K)$ is unknown.

The equality between $V_\Sigma(K)$ and $V^*(K)$ holds if $V_\Sigma(K)$ is closed. From examples we have seen that it is possible that $V_\Sigma(K)$ is not closed, but this property is realisation dependent. This motivates the following (still) open problem: under what conditions does the transfer function $f(s)$ have a realisation (D,A,B) such that $f(s) = D(s-A)^{-1}B$ and $V_\Sigma(Ker\, D)$ is closed. In order to solve this one would probably need the concepts of open loop invariance, closed loop invariance and frequency invariance for more general systems, for instance with an unbounded input operator B. However there is reason to believe that the following conjecture will hold; let (A,B) be a general systems with possibly unbounded input operator B, and let the feedback laws F be elements of some class of linear operators such that $A+BF$ generates a C_0-semigroup, then the largest closed loop invariant subspace, with respect to our class of feedback operators, does not necessarily exist.

So geometric theory for infinite dimensional systems has brought and will continue to produce some nice mathematical and system theoretic results, but it will be very hard to apply these results to applications.

REFERENCES

1. G. BASILE and G. MARRO;

    Controlled and Conditioned Invariant Subspaces in Linear System Theory. Journal of Optimization Theory and Applications, Vol. 3, no. 5, pp. 306–315, 1969.

2. G. BASILE, G. MARRO and A. PIAZZI;

    Revisiting the Regulator Problem in the Geometric Approach, Part 1 Disturbance Localization by Dynamic Compensation, Part 2, Asymptotic Tracking and Regulation in the Presence of Disturbances. Journal of Optimization Theory and Applications, Vol. 53, no. 1, pp. 9–36, 1987.

3. R. BELLMAN and K.L. COOKE;

    Differential–Difference Equations, Academic Press, New York, 1963.

4. B.M.N. CLARKE and W.F. HOLLAND;

    Eigenstructure Specification for Linear Systems in Hilbert Space I Spectral scalar operators. Macquarie Mathematics Reports, no. 85–0060, School of Mathematics and Physics, Macquarie University, 1985.

5. R.F. CURTAIN;

    Spectral Systems. Int. J. Control, 39, pp. 657-666, 1984.

6. R.F. CURTAIN;

    Invariance Concepts in Infinite Dimensions. SIAM Journal of Control and Optimiz, Vol. 24, no. 5, pp. 1009–1031, Sept. 1986.

7. R.F. CURTAIN;

    (C,A,B)–pairs in Infinite Dimensions. System and Control Letters, 5, pp. 59–65, 1984.

8. R.F. CURTAIN;

    Disturbance Decoupling by Measurement Feedback with Stability for Infinite Dimensional Systems. Int. J. Control, Vol. 43, no. 6, pp. 1723–1743, 1986.

9. R.F. CURTAIN and A.J. PRITCHARD;

    Infinite Dimensional Linear Systems Theory. Lecture Notes in Control and Information Sciences 8, Springer Verlag, Berlin, 1978.

10. R.F. CURTAIN and H.J. ZWART;

   Spectral Realisation for Delay Systems. Distributed Parameter Systems, Proceedings of the 3rd International Conference on Control of Distributed Parameter Systems, Eds. F. Kappel, K. Kunisch, W. Schappacher, July 6–12, Vorau, Lecture Notes in Control and Information Sciences 102, Springer Verlag, Berlin, pp. 64–89, 1986.

11. R.F. CURTAIN and H.J. ZWART;

   L_∞–approximations of Nonrational Transfer Functions: An example. pp. 167–168, Proc. of the 25th IEEE Conference on Decision and Control, Dec. 10–12, 1986, Athens, Greece, IEEE Control Systems Society, New York, 1986.

12. R.F. CURTAIN and H.J. ZWART;

   A note on Spectral Realisations for Delay Systems. to appear in Systems and Control Letters, 1989.

13. R. DATKO;

   A linear Control Problem in a Abstract Hilbert Space. Journal of Differential Equations 9, pp. 346–359, 1971.

14. E.B. DAVIES;

   One Parameter Semigroups. Academic Press, London, 1980.

15. E.J. DAVISON and S.J. WANG;

   Properties and Calculation of Transmission Zeros of Linear Multivariable Systems. Automatica, Vol. 10, pp. 634–658, 1974.

16. G. DOETSCH;

   Introduction to the Theory and Application of the Laplace Transformation. Springer verlag, Berlin, 1974.

17. N. DUNFORD and J.T. SCHWARTZ;

   Linear Operators Part I: General Theory. Interscience Publishers Inc., 1958.

18. N. DUNFORD and J.T. SCHWARTZ;

   Linear Operators, Part III: Spectral Operators. Wiley-Interscience, New-York, 1971.

19. M.L.J. HAUTUS;

   (A,B)–invariant and Stabilizability Subspace, a Frequency Domain Description. Automatica, Vol. 16, pp. 703–707, 1980.

20. M.L.J. HAUTUS;

> Controlled Invariance in Systems over Rings. Memorandum COSOR 82–01, Department of Mathematics and Computer Science, Eindhoven University of Technology, The Netherlands, 1982.

21. C.A. JACOBSON and C.N. NETT;

> Linear State–Space Systems in Infinite–Dimensional Space: The Role and Characterization of Joint Stabilizability/Detectability. IEEE Trans. Automat. Control., Vol. AC–33, pp. 541–549, 1988.

22. T. KATO;

> Perturbation Theory for Linear Operators. Springer Verlag, Berlin, 1966.

23. T.G. KURTZ;

> A General Theorem on the Convergence of Operator Semigroups. Transactions of the American Mathematical Society, 148, pp. 23–32, March 1980.

24. I. LASIECKA and R. TRIGGIANI;

> Finite Rank, Relatively Bounded Perturbations of Semigroups Generators, Part I. Annali Scuola Normale Superiore–Pisa, Classe di Scienze, Serie IV – Vol. XII, no. 4, 1985.

25. S.A. NEFEDOV and F.A. SHOLOKHOVICH;

> A Criterium for the Stabilizability of Dynamical Systems with Finite–Dimensional Input. Differential Equations: Translations of Diffentsial'nye Uravneniya, pp. 163–166, New York, Plenum, 1986.

26. L. NOOITGEDAGT;

> Computation of Transmission Zeros for Distributed Parameter Systems and an Application to Spectral Systems. Master of Science Thesis, Department of Mathematics, University of Groningen, Groningen, March, 1986.

27. L. PANDOLFI;

> Disturbance Decoupling and Invariant Subspaces for Delay Systems. Applied Mathematics and Optimization 14, pp. 55–72, 1986.

28. J.R. PARTINGTON, K. GLOVER, H.J. ZWART and R.F. CURTAIN;

> L_∞–approximation and Nuclearity of Delay Systems. Systems and Control Letters 10, pp. 59–65, 1988.

29. A. PAZY;

> Semigroups of Linear Operators and Applications to Partial Differential Equations. Springer Verlag, New York, 1983.

30. S. POHJOLAINEN;

Computation of Transmission Zeros for Distributed Parameter Systems. Report 34, Department of electrical Engineering, Mathematics, Tampere University of Technology, Tampere 1979.

31. W. RUDIN;

Real and Complex Analysis. McGraw–Hill Book Company, New-York, 1966.

32. E.J.P.G. SCHMIDT and R.J. STERN;

Invariance Theory for Infinite Dimensional Linear Control Systems. Applied Mathematics and Optimization 6, pp. 113–122, 1980.

33. J.M. SCHUMACHER;

Algebraic Characterizations of Almost Invariance. Int. J. Control, Vol. 38, no. 1, pp. 107–124, 1983.

34. J.M. SCHUMACHER;

Compensator Synthesis using (C,A,B)–pairs. IEEE Trans. on Automatic Control, AC–25, pp. 1133–1138, 1980.

35. J.M. SCHUMACHER;

Dynamic Feedback in Finite and Infinite–Dimensional Linear Systems. Mathematical Centre Tracts No. 143, Mathematisch Centrum, Amsterdam, 1982.

36. R.J. STERN;

Asymptotic holdability. IEEE Transactions on Automatic Control, vol. AC–25 no. 6, 1980.

37. S.H. SUN;

On Spectrum Distribution of Completely Controllable Linear Systems. Acta Mathematica Sinica, 21, pp. 193–205, 1978. translation by Ho, L.F., SIAM J. Control and Optimization, 19, no. 6, pp. 730–743, 1981.

38. T. TAKAMATSU, I. HASHIMOTO and Y. NAKAI;

A Geometric Approach to Multivariable Control System Design of a Distilation Column. Proc. 7'th Triennial World Congress IFAC, Helsinki, pp. 309–317, 1978.

39. H.L. TRENTELMAN;

Almost Invariant Subspaces and High Gain Feedback. C.W.I. Tract 29, Centre for Mathematics and Computer Science, Amsterdam, 1987.

40. P. WERMER;

Commuting Spectral Operations on Hilbert Spaces. Pacific J. Math., 4 pp. 355–361, 1954.

41. J.C. WILLEMS and C. COMMAULT;

Disturbance Decoupling by Measurement Feedback with Stability or Pole Placement. SIAM J. Control and Optimiz, 19, pp 490–504, 1981.

42. W.M. WONHAM;

Linear Multivariable Control: A Geometric Approach. Springer Verlag, A.M.S. 10, 1978.

43. J. v.d. WOUDE;

Feedback Decoupling and Stabilization for Linear Systems with Multiple Exogenous Variables. Ph. D. Thesis, University of Eindhoven, 1987.

44. K. YOSIDA;

Functional Analysis. Grundlehren der Mathematischen Wissenschaften 123, Springer Verlag, Berlin 1978.

45. J. ZABCZYK;

On Decomposition of Generators. SIAM J. Control and Optimization, Vol. 16, no. 4, pp. 523–534, July 1978.

46. A.H. ZEMANIAN;

Distribution Theory and Transform Analysis, An Introduction to Generalized Functions, with Applications. Mc. Graw–Hill Book Company, 1965.

47. H.J. ZWART;

Characterization of all Controlled Invariant Subspaces for Spectral Systems. SIAM Journal of Control and Optimization, Vol. 26, no. 2, pp. 369–387, 1988.

48. H.J. ZWART;

Equivalence between Open Loop and Closed Loop Invariance for Infinite Dimensional Systems: A Frequency Domain Approach. SIAM Journal of Control and Optimization, Vol. 26, no. 5, 1988.

49. H.J. ZWART;

On Stabilizability of Infinite Dimensional Systems. preprint, 1988.

50. H.J. ZWART and J. BONTSEMA;

Spectral Systems. preprint, 1988.

51. H.J. ZWART, R.F. CURTAIN, J.R. PARTINGTON and K. GLOVER;

Partial Fraction Expansions for Delay Systems. Systems and Control Letters 10, pp. 235–243, 1988.

52. H.J. ZWART and L. NOOITGEDAGT;

Zeros for Discrete Spectral Systems with an application to the Disturbance Decoupling Problem. Rapport TW-272, Mathematisch Instituut, Rijksuniversiteit Groningen, 1986.

LIST OF ALL INVARIANCE CONCEPTS AND THEIR RELATION

Definitions:

A–invariant	$A(V \cap D(A)) \subset V$.
(A, B)–invariant	$A(V \cap D(A)) \subset V + Im\, B$.
closed loop invariant	there exists a bounded F such that $T_{A+BF}(t)V \subset V$ $\forall t \geq 0$.
conditioned invariant	there exists a bounded G such that $T_{A+GC}(t)V \subset V$ $\forall t \geq 0$.
controlled invariant	closed loop invariant = open loop invariant = frequency invariant for closed subspaces V.
feedback invariant	there exists an A–bounded F such that $(A+BF)(V \cap D(A)) \subset V$.
frequency invariant	For every $x_0 \in V$ there exist $\xi(.) \in D(A)(s)$ and $\omega(.) \in U_{-1}(s)$ such that $x_0 = (s-A)\xi(s) - B\omega(s)$.
generator invariant	A–invariant.
open loop invariant	For every $x_0 \in V$ there exists a continuous $u(.)$ such that the solution of $\dot{x}(.) = Ax(.) + Bu(.)$; $x(0) = x_0$ remains in V.
semigroup invariant	$T_A(t)$–invariant.
$T_A(t)$–invariant	$T_A(t)V \subset V$; $\forall t \geq 0$.

Relations:

For a <u>closed</u> subspace V we have that the following concepts are equivalent:

i)	closed loop invariance.
ii)	open loop invariance.
iii)	frequency invariance.
iv)	controlled invariance.

Furthermore the following two concepts are also equivalent for a <u>closed</u> subspace V.

v)	(A, B)–invariance.
vi)	feedback invariance.

Note: For a closed subspace V i) always implies v), but the converse does not hold in general.

A	The system operator, which is assumed to generate the C_0–semigroup $T_A(t)$.
B	The input operator, which is assumed to be bounded and to have finite dimensional range.sional range.
$B^0(V)$	A subspace of $Im\,B$ which has zero intersection with V and is of maximal dimension under this restriction. restriction.
$B^1(V)$	$Im\,B \cap V$.
C	An output operator, which is assumed to be bounded.
D	An output operator, which is assumed to be bounded.
$D(A)$	The domain of the operator A.
E	The disturbance input operator, which is assumed to be bounded.
F	A feedback operator from the state space to the input space.
G	A feedback operator from an output space to the state space.
$H^2(V)$	The space of Hardy functions with values in V. V is assumed to be a closed subspace of a Hilbert space.
$\mathcal{H}$	The state space, if this state space is a Hilbert space.
i_V	The inclusion map, i.e. if $V \subset \mathcal{X}$, then $i_V : V \mapsto \mathcal{X}$ and $i_V(x) = x$.
$Im\,B$	The image of the operator B.
$\mathcal{I}m(z)$	The imaginary part of the complex number z.
K	This is the feed through term in the compensator, or an arbitrary closed subset of the state space. The meaning will be clear from the context.
$Ker\,D$	The kernel of the operator D.
L	The output operator from the compensator, or the distribution diagram (only in appendix E).
$L^2([a,b);Q)$	The space of all square integrable functions from $[a,b)$ to Q.
$\mathcal{L}$	The Laplace transform.
$\mathcal{L}(\mathcal{X})$	The space consisting of all bounded linear operators from $\mathcal{X}$ to $\mathcal{X}$.
$\mathcal{L}(\mathcal{X},\mathcal{Z})$	The space consisting of all bounded linear operators from $\mathcal{X}$ to $\mathcal{Z}$.
M	The input operator in the compensator.

N	The system operator for the compensator, which is assumed to generate a C_0–semigroup.
(N,M,L,K)	The compensator.
P	A projection operator.
P_V	The projection operator on V.
Q	An arbitrary operator.
$\mathcal{Q}$	The disturbance input space.
$Re(z)$	The real part of the complex number z.
S	A subspace of the state space.
$S^*(Im\,E)$	The smallest conditioned invariant subspace that containes $Im\,E$
$T_A(t)$	The semigroup generated by A.
$\mathcal{U}$	The controlled input space.
V	A subspace of the state space.
$V^*(K)$	The largest controlled invariant subspace contained in the closed subspace K.
$V_{ol}(K)$	The largest open loop invariant subspace contained in the closed subspace K.
$V_\Sigma(K)$	The largest frequency invariant subspace contained in the closed subspace K.
$\mathcal{V}_{(A,B)}(K)$	$V_\Sigma(K)$ for the system (A,B).
W	The state space of the compensator.
$\mathcal{X}$	The state space of the system if this state space is an arbitrary Banach space.
$\mathcal{Y}$	The output space.
$Y(s)$	The space consisting of all function $f(.)$ from $\mathbb{R}$ to Y which are continuous on some interval of the from $[r,\infty)$.
$Y_{-1}(s)$	$\{f\in Y(s)\mid \underset{s\to\infty}{lim}\; sf(s)\text{ exists}\}$
$\mathcal{Z}$	The output space of the to be controlled output.
Ξ_s	All states that can be reached at frequency s, under the condition that the state trajectory has to stay in V.
Ξ	Ξ_s for s sufficiently large.
$*$	The $*$ will denote the Hilbert space dual of an operator.
	The $'$ will denote the Banach space dual.
$\mathbf{1}_{[a,b]}$	The indicator function of the interval $[a,b]$.

INDEX

Lecture Notes in Control and Information Sciences

Edited by M. Thoma and A. Wyner

Lecture Notes in Control and Information Sciences

Edited by M. Thoma and A. Wyner

Lecture Notes in Control and Information Sciences

Edited by M. Thoma and A. Wyner